"Ovaska and Brei composed their unique text around a trinity: moral theories, moral self-licensing, and day-to-day cases from the professional life of engineers. They clearly know their stuff. As a professional with long career in industry, academia and government, I still learned so much from the book. This is a must read for every engineer, scientist and technology professional."

Phil Laplante, *Emeritus Professor of Software and Systems Engineering, Penn State*

Ethics for Engineers

Ethics for Engineers: Toward Ethical Behavior within Engineering Organizations offers a multilevel perspective on engineering ethics with considerable breadth and depth, making it a valuable resource for students, educators, and professionals alike.

This pragmatic book contains case studies of micro-level ethical violations, evaluating their moral implications and discussing moral self-licensing behind making unethical decisions. It also explores macro-level cases that have caused significant reputational and financial damage to major companies. In addition, the authors touch on topics whose overall impact is not yet fully understood, such as environmental ethics issues related to wind turbine blades and space debris management. By presenting examples from different levels and offering reflections from various perspectives, this text prompts readers to critically evaluate the ethical implications of their actions and understand what may drive a work community to behave unethically.

Key features:

- Covers both moral theoretical and behavioral ethics perspectives.
- Contains day-to-day micro-level cases from the lives of practicing engineers, supplemented with macro-level cases.
- Provides pragmatic guidance for individual engineers and their organizations to move toward value-based ethics.
- Features colloquial language to make the book an enjoyable and accessible read.
- Includes 29 demonstrative vignettes, 87 class exercises, and an insightful interview with an ethics ambassador.

This unique text serves as a pedagogically sound learning companion for courses in engineering ethics and related topics, striking a balance between research-based findings (with over 40 scholarly references) and real-world experiences (featuring an Appendix by an industry executive).

What Every Engineer Should Know

Series Editor Phillip A. Laplante, Pennsylvania State University

What Every Engineer Should Know about Cyber Security and Digital Forensics
Joanna F. DeFranco

What Every Engineer Should Know about Modeling and Simulation
Raymond J. Madachy and Daniel Houston

What Every Engineer Should Know about Excel, Second Edition
J.P. Holman and Blake K. Holman

Technical Writing: A Practical Guide for Engineers, Scientists, and Nontechnical Professionals, Second Edition
Phillip A. Laplante

What Every Engineer Should Know About the Internet of Things
Joanna F. DeFranco and Mohamad Kassab

What Every Engineer Should Know about Software Engineering
Phillip A. Laplante and Mohamad Kassab

What Every Engineer Should Know About Cyber Security and Digital Forensics
Joanna F. DeFranco and Bob Maley

Ethical Engineering: A Practical Guide with Case Studies
Eugene Schlossberger

What Every Engineer Should Know About Data-Driven Analytics
Satish Mahadevan Srinivasan and Phillip A. Laplante

What Every Engineer Should Know About Reliability and Risk Analysis
Mohammad Modarres and Katrina Groth

What Every Engineer Should Know About Smart Cities
Valdemar Vicente Graciano and Mohamad Kassab

What Every Engineer Should Know About Digital Accessibility
Sarah Horton and David Sloan

What Every Engineer Should Know About Python
Raymond J. Madachy

Diversity in STEM: Analyzing Inequities and Improving Opportunities in Education and the Workplace
Benjamin Blocker II

Ethics for Engineers: Toward Ethical Behavior within Engineering Organizations
Seppo J. Ovaska and Andrew T. Brei

For more information about this series, please visit: www.routledge.com/What-Every-Engineer-Should-Know/book-series/CRCWEESK

Ethics for Engineers

Toward Ethical Behavior within Engineering Organizations

Seppo J. Ovaska and Andrew T. Brei

CRC Press is an imprint of the
Taylor & Francis Group, an **informa** business

Designed cover image: Seppo J. Ovaska

First edition published 2026
by CRC Press
2385 NW Executive Center Drive, Suite 320, Boca Raton, FL 33431

and by CRC Press
4 Park Square, Milton Park, Abingdon, Oxon, OX14 4RN

CRC Press is an imprint of Taylor & Francis Group, LLC

ISBN: 978-1-032-77943-0 (hbk)
ISBN: 978-1-032-77942-3 (pbk)
ISBN: 978-1-003-48552-0 (ebk)

DOI: 10.1201/9781003485520

Typeset in Times
by KnowledgeWorks Global Ltd.

Seppo:

To Helena, Sami and Samu—with all my love

Andrew:

To Kim, who makes everything around her better… including me

Contents

Foreword

The importance and prevalence of technology in our daily lives have increased dramatically in recent decades. Although the field of information technology has faced the most rapid growth, groundbreaking innovations have also emerged from diverse areas, such as medicine, biotechnology, and materials science. Without the efforts of engineers and their inventions, our world would be a vastly different place.

However, in parallel with the stunning technological advancements, the humankind has faced a series of colossal threats that require the expertise of engineers to overcome. Whether it is finding solutions to mitigate the climate change or designing machinery to clean up the enormous garbage patches in our oceans, the power of engineers has probably never been greater. But with such a great power comes great responsibility, and it is hence even more important for engineers to consider the ethical implications of their work. The solutions of engineers definitely shape our planet, and thus examining the ethical dimension of engineering is an essential skill and also a topic whose significance may not always be apparent for young professionals.

It must be stressed that bad things are often made by good people. Although engineers are often revered as problem-solvers and innovators, they are also human beings who can experience immense pressure in their work. Whether it is the pressure of meeting employer-defined goals, the threat of impending project funding cuts, or the societal fear of replacing people with machines using AI technology, engineers can and will find themselves in stressful situations during their career. The pressure can also be intrinsic and, sometimes, not even concrete to anyone but oneself. When facing intense pressure, it is not simple to maintain a clear perspective and refrain from making unethical decisions, either consciously or not.

In recent times, there has been a growing emphasis on ethics and compliance with codes of conduct, as well as on fair and just culture in work communities. Previously in the working life, an ethical violation often meant for the majority a large-scale systematic fraud or a serious infringement of employees' rights, but attention is now being paid also to small-scale unethical actions. From a philosophical perspective, both large and small unethical actions are unacceptable and wrong, although ethical violations can be placed in a hierarchy based on how much they violate certain core virtues. In order to foster a culture of integrity and responsibility, it is crucial to avoid unethical actions regardless of their scale; and by addressing small unethical actions effectively, we can potentially reduce the number of major unethical actions, too. While having a moral compass guiding us toward making ethical decisions, anyone can practice the skills of making ethical decisions by considering the diverse consequences of their actions.

Seppo J. Ovaska and Andrew T. Brei's textbook, *Ethics for Engineers: Toward Ethical Behavior within Engineering Organizations*, offers a unique multilevel perspective on engineering ethics. Their book examines case studies of day-to-day micro-level ethical violations, evaluating their moral implications and discussing moral self-licensing behind making a bad/wrong decision; as well as explores serious, large-scale cases that have caused, for instance, significant reputational and financial damage to major companies. The authors also touch on topics whose impact on the planet or the universe is not yet fully understood, such as environmental problems related to wind energy production and poor waste management of objects in space. By presenting examples from different levels and providing moral theoretical explanations and reflections from various perspectives, this book prompts readers to evaluate the ethical implications of their own everyday actions and to understand what may drive an entire work community to behave unethically.

It is my pleasure to recommend this book as a pragmatic guide for engineers seeking to navigate the increasingly complex ethical challenges of the modern world. The book takes a unique approach to examining ethics for engineers through diverse case studies and philosophical reflection. It is certainly to be of significant and long-lasting value to students, young professionals, researchers, and even senior engineers in different fields of engineering.

Sami-Seppo Ovaska
Pulp and Paper Engineer, Ethics Ambassador
Stora Enso, Finland

Preface

INTRODUCTION

Ethics for Engineers is a diverse field that could be presented in many different ways. The subtitle, *Toward Ethical Behavior within Engineering Organizations*, reflects the viewpoint of the present treatment. This book explores many of the ethical issues that an engineer faces in the course of his/her professional engineering practice. Some engineering ethics textbooks focus on big unethical cases, such as the NASA Space Shuttle accidents, the toxic-gas disaster in Bhopal, India, or the more recent diesel emissions scandal of Volkswagen automobiles. But our unique text has its focus on the perspective of *individual actors* in small, everyday cases, and their immediate consequences for the actors themselves. Our hypothesis is that as engineers learn to control their smaller ethical challenges, also the number of bigger unethical cases where they would be involved gradually decreases.

Sometimes actors can also be seen as victims of their own unethical behavior, such as dishonesty. Unethical behavior may come at an internal cost by increasing tension and harming the unethical person's moral self-image; this syndrome is called an *ethical hangover*. We recognize an analogy between the ethical hangover and the alcohol hangover that results from drinking too much. Hence, ethical hangovers could naturally be prevented by behaving "not too unethically." But the concept of ethical behavior is dual-level for engineers working in companies. Due to the obligations imposed by their employer, practitioners must also consider the ethical standards of the organization—even if they violate their personal ethical standards.

After an introduction to three famous *moral theories*, including the Utilitarian theory of Jeremy Bentham, the Deontological theory of Immanuel Kant, and the Virtue Theory of Aristotle, we basically have theoretical tools available for distinguishing what is right and what is wrong. However, it is apparent that these tools have limitations in analyzing everyday cases in engineering organizations. Still, because intentions, decisions, actions, and values are part of everyone's life, morality is an inescapable—for engineers, as well. Furthermore, existing professional ethical codes, such as the IEEE Code of Ethics, have a moral theoretical basis.

Next, we stray away from the moral theories to behavioral ethics and adopt a complementary tool, *moral self-licensing*, for explaining the behavior of engineers in making unethical decisions and performing corresponding actions. Moral self-licensing occurs when evidence of a person's virtue frees him/her to act less-than-virtuously. This kind of moral behavior involves equating being a good engineer with striking a balance between right and wrong actions such that, overall, the scale is tipped toward the good. Behavioral ethics is an applied field of study based on behavioral psychology, cognitive science, and evolutionary biology—rather than moral philosophy.

Carefully selected, day-to-day cases are presented and analyzed/explained with moral theoretical and moral self-licensing tools. The unethical cases, studied in detail, have their roots in the real-world, but the authors have adopted a certain authors' license in telling the stories, and thus preserving the anonymity of individuals and companies involved. These micro-level case studies consist of common and frequently emerging themes that can be classified into three categories: *product*, *employment*, and *interaction*. Where relevant, we speculate on how such cases would be handled in organizations with no established ethical culture and, on the other hand, in organizations with advanced ethical culture. We also highlight the overall consequences to the actors themselves. The key virtues we pay attention to in our analysis are honesty, integrity, respect, responsibility, and self-discipline. In addition, a few macro-level cases are studied from the perspective of temporal and spatial consequences.

AUDIENCE

This book is an introductory text about ethics for engineers, and it is intended for junior–senior level students and young professionals. And it is also recommended for more established engineers and engineering managers who want to absorb value-based ethics as a new avenue in practicing their profession. Moreover, it may find relevance in *Organizational Behavior* courses for business and management students, too. Therefore, the potential pool of people who could benefit from our book is considerable. We expect this textbook to have a long sales life, and its main markets are in Europe and North America.

COURSE ADOPTION

Academic course titles where our book could be used as a primary or complementary text were searched from the web, and they include the following examples:

- Engineering and Ethics
- Engineering Ethics
- Engineering Ethics and Ethical Choices
- Engineering Ethics and Its Impact on Society
- Ethics for Engineers
- Professional Ethics in Engineering

With genuinely recognized ethical values coupled with exemplary leadership, the entire engineering profession would be a more attractive and friendlier environment for prospective professionals to join and work. We do hope that this textbook contributes in its part to achieving that noble goal.

"All I've done is give you a book. You have to have the courage to learn what's inside it." These words of encouragement were offered by science teacher Freida Joy Riley to a high-school student back in the late 1950s (H. H. Hickam, Jr., *October Sky: A Memoir*. New York, NY: Dell Publishing, 1999, p. 232). May her enduring words also inspire the potential readership of our book.

Let's do what's right!

ERROR HANDLING

Where verbatim phrases were used, the authors tried to cite them appropriately. However, if any were inadvertently overlooked, the authors wish to correct the unfortunate errors. Please notify the authors if you find any errors of omission, citation, and so forth by e-mail, at *abrei@stmarytx.edu*, and they will be corrected at the next possible opportunity.

Seppo J. Ovaska
Hyvinkää, Finland

Andrew T. Brei
San Antonio, Texas

Acknowledgments

Seppo Ovaska wishes to thank his dear friend Dr. Andrew Brei for being the perfect collaborator. Easy to work with, Andrew's insight, attention to moral details, and the philosopher's viewpoint complemented Seppo's strengths and weaknesses. "Andrew, it was a true pleasure to work with you in this exciting and rewarding book project."

Seppo also wishes to thank his wife Helena and his sons Sami and Samu for putting up with the seemingly endless work on this manuscript and many other writing projects over the years. "Although it is a tiny gesture compared with all that you have given to me, I humbly dedicate this book to you."

And finally, Seppo wants to express heartfelt gratitude to his teachers: Kirsti Länsisyrjä (elementary school), Matti Tarsaranta (high school), and Dr. Yrjö Neuvo (university). "Your unwavering dedication, patience, and wise guidance, along with your exemplary role modeling, have profoundly influenced my growth. You were prodigious!"

Andrew Brei has immense gratitude for Dr. Seppo Ovaska's kindness, energy, collegiality, encouragement, and patience throughout this project. "Seppo, you've been a most supportive collaborator, and I value your friendship and support."

Andrew would also like to thank the faculty members at St. Mary's University who have made working at the intersections of philosophy and engineering so rewarding. These valued colleagues include Morgan Bruns, Gopal Easwaran, Wenbin Luo, Matthew Mangum, Rafael Moras, Juan Ocampo, and Robert Boyd Skipper. Thanks also to Marie Peterson, Research Assistant for the Center for Professional Ethics at St. Mary's University, for her honesty, insight, and consideration.

Finally, Andrew wishes to thank his partner, Kim, and daughters, Ella and Mia. His tendencies toward procrastinating, complaining, and daydreaming were no match for their faithful support and reliable encouragement.

The authors are grateful to the field expert reviewers, Professor Gregory D. Buckner (North Carolina State University), Professor Todd K. Moon (Utah State University), and Professor Emeritus Robert Boyd Skipper (St. Mary's University), for their valuable time and truly significant efforts aimed at improving the manuscript.

S.J.O. & A.T.B.

About the Authors

Dr. Seppo J. Ovaska received his D.Sc. degree from Tampere University of Technology in 1989. He served as a Professor at the Aalto University School of Electrical Engineering and held a Visiting Professorship at Utah State University in 2006–2007. Dr. Ovaska retired in 2019 after 27 years in university faculty and 13 years in industrial R&D—in Finland and the United States. For four decades, embedded systems were at the heart of his engineering and teaching endeavors. His research expertise includes computational intelligence, green computing, and industrial electronics. Professor Ovaska delivered his farewell lecture, "Cybersecurity and Machine Learning," at the University of Kassel in Germany. He has (co-)authored over 270 scholarly papers and is the (co-)inventor of six U.S. patents. This is his fourth book. Currently, Dr. Ovaska is interested in moral philosophy and moral self-licensing in the context of engineering ethics. He is a *First Class Knight of the Order of the White Rose of Finland.*

Dr. Andrew T. Brei earned Bachelor's degrees in German and philosophy from the University of Wisconsin – Stevens Point in 1996 and a Doctorate in philosophy from Purdue University in 2009. Since 2010, he has been part of the Philosophy Department at St. Mary's University. In his teaching and scholarship, Dr. Brei investigates a wide range of topics, including moral motivation, environmental philosophy, engineering ethics, human nature, the philosophy of food, the nature of reality and knowledge, the moral status of bullfighting, and the ethics of extra-terrestrial exploration. In 2019, Dr. Brei earned the *Distinguished Faculty Award for Excellence in Teaching* from the St. Mary's University Alumni Association. He has delivered webinars and keynote addresses to various professional organizations by way of promoting and demystifying ethical reasoning. Currently, Dr. Brei is exploring the nature and implications of travel in an emerging branch of philosophy called, appropriately enough, the Philosophy of Travel.

About the Authors

Acronyms

AI	Artificial Intelligence
AQVM	Actor's Qualitative Virtue Measure
BMW	Bayerische Motoren Werke
CAD	Computer-Aided Design
CD	Compact Disc
CEO	Chief Executive Officer
CEV	Corporate Ethical Virtues
CPU	Central Processing Unit
CI	Categorial Imperative
CSDDD	Corporate Sustainability Due Diligence Directive
CSRD	Corporate Sustainability Reporting Directive
CTO	Chief Technology Officer
CV	Curriculum Vitae
DSP	Digital Signal Processing
EPA	Environmental Protection Agency
EPO	European Patent Office
ESA	European Space Agency
ESG	Environmental, Social and Governance
EU	European Union
EUDD	European Union Deforestation Directive
EV	Electric Vehicle
GEO	Geostationary Equatorial Orbit
GPA	Grade Point Average
HDTV	High-Definition Television
HIT	Honesty, Integrity and Truthfulness
HR	Human Resources
HVAC	Heating, Ventilation and Air Conditioning
ICT	Information and Communications Technology
IEEE	Institute of Electrical and Electronics Engineers
IP	Intellectual Property
IPR	Intellectual Property Rights
ISO	International Organization for Standardization
IT	Information Technology
KPI	Key Performance Indicator
LEO	Low Earth Orbit
LGBTQ+	Lesbian, Gay, Bisexual, Transgender, Questioning and more
MEO	Medium Earth Orbit
MS	Multiple Sclerosis
NASA	National Aeronautics and Space Administration
NIAAA	National Institute on Alcohol Abuse and Alcoholism
NSPE	National Society of Professional Engineers
OOP	Object-Oriented Programming

PD	Product Development
PH	Principle of Humanity
PU	Principle of Universalizability
R&D	Research and Development
SLC	Salt Lake City
SSR	Soviet Socialist Republic
UML	Universal Modeling Language
UN	United Nations
VW	Volkswagen
WHO	World Health Organization
WNA	World Nuclear Association
3D	3-Dimensional

1 Ethics for Practicing Engineers

Learning Objectives

After studying Chapter 1, you will be able to

- Recall the basic terminology related to ethics for engineers
- Understand the progressive chain between workplace ethics, work atmosphere, personal wellbeing, and organizational effectiveness
- Recognize the importance of ethics in engineering organizations and thus to your career as a practicing engineer

Engineering ethics is an undervalued issue among engineering professionals. Although most engineers agree that general ethical principles should be followed in everyday life, it is quite rare that the principles of ethics are explicitly on the table when engineers make decisions and take corresponding actions. And yet, whenever decisions are made or actions are taken, morality is always involved.

In this introductory text, an *engineer* is any professional with a degree in engineering; regardless of whether he/she is engaged in engineering or management work—"once an engineer, always an engineer." And the concept of *ethics* can be succinctly defined as the study of morality—or, more precisely, the field of inquiry which addresses questions about what our guiding ideals should be, what sort of life is worth living, and how we should treat one another [1]. Furthermore, *engineering ethics* is a segment of moral philosophy that shows the ways in which engineers should behave and act in their professional capacity, at large. This also includes interaction with their colleagues and stakeholders.

A concern for ethics is now on the upswing in many industrial workplaces; at least in Europe and North America, as well as in Australia. More and more companies today are becoming aware of ethical issues; mere compliance with laws is no longer regarded as good enough, and the importance of ethical values in advanced organizations is growing. The slogan "do what's right" is becoming increasingly appreciated. This requires genuine commitment from the management—leading by example—and persistent investments in personnel training. As an example, the Ethics Ambassador Program of Stora Enso, a leading provider of renewable products in packaging, biomaterials, and wooden construction, is a promising step in this direction [2]. And it is, indeed, easy to agree with the statement of Kim et al. [3]: "We cannot emphasize enough the importance of engineering ethics education throughout college and into workplaces to shape a foundation for further ethical development among engineers." Although this textbook is aimed at junior–senior

DOI: 10.1201/9781003485520-1

courses in engineering ethics, it may also find relevance in organizational behavior courses for business and management students.

Before we continue our mission to spread the happy message of engineering ethics, it is useful to look at a sample of unethical actions in real industrial organizations to get an idea of *what we are aiming to avoid* by behaving more ethically. The following actions of fraud and misconduct in the workplace were picked from an integrity survey of KPMG Forensic [4]:

a. Engaging in false or deceptive sales practices
b. Accepting inappropriate gifts or kickbacks from suppliers
c. Falsifying time and expense reports
d. Engaging in fraudulent or illegal acts
e. Violating environmental standards
f. Engaging in sexual harassment or creating a hostile work environment

This representative shortlist is compiled from more than 40 reported categories of unethical actions in American industry. And based on our international engineering experience, we can confidently say that similar problems exist worldwide. Hence, there is a wide variety of options for unethical behavior in the workplace; it is a multifaceted problem.

WHAT IS THIS ETHICS STUFF ALL ABOUT?

From Virtues to Values

Engineering ethics consists of guidelines, principles, and practices for engineers to follow to ensure their decision-making and actions are aligned with their explicit and implicit obligations to the public, their customers, the shareholders, as well as their organization and colleagues. An intuitive framework of engineering ethics can be built around a number of virtues that individual engineers practice to a greater or lesser extent (Table 1.1)—hence "an ethical engineer is a virtuous engineer."

But what is a *virtue*? A virtue is a character trait—a disposition to act in a particular positive way. Solomon defined virtue as: "An exemplary way of getting along with other people, a way of manifesting in one's own thoughts, feelings and actions the ideals and aims of the entire community" [5]. A common feature of virtues is that they must be acquired through practice. Another feature is that they have, as it were, two opposites [6]. Take *honesty*, for instance. It is a virtue, and it corresponds to vices called *bluntness* at one extreme end and *dishonesty* at the other. Therefore, it should be noted that a character trait is only good—and, thus, worthy of the name virtue—if it is not pushed to either extreme. When it is so pushed, it becomes a different trait; it becomes a *vice*. This is what Aristotle meant when he described virtue as "… a mean between two vices, one of excess and one of deficiency" [7]. Thus, being an ethical person includes nurturing and acting on virtues and avoiding vices.

TABLE 1.1
A Representative Set of Virtues Practiced in Ethical Workplaces

Virtue
Determination
Fairness
Honesty
Integrity
Loyalty
Respect
Responsibility
Self-discipline
Tact
Tolerance
Trustworthiness
Truthfulness

We recommend that the class first defines the virtues listed in Table 1.1 in small groups, and then as the whole class discusses the definitions. For example, are *honesty* and *truthfulness* synonymous? No, there is a subtle difference between them: honesty means being sincere and non-deceptive, while truthfulness means being strictly accurate and factual. We will provide a deeper discussion of virtues and other important moral considerations in Chapter 3.

Another core concept, *value*, serves as a criterion and standard for our behavior. Values guide how we evaluate or choose our actions, events, habits, and even other people. We decide what is good or bad, right or wrong, worth doing or avoiding, based on the personal values we hold. Our values often reflect the ways of living we find around us, involving family, friends, and the wider community. So, it is vital that, from time to time, every individual reflects on and evaluates their values to avoid being a "puppet," controlled by what others think is important and worthy. Furthermore, people tend to act in ways that allow them to express their important values and to achieve the goals that underlie them [8]. Finegan studied the impact of personal values on judgments of ethical behavior in the workplace, and she found that the personal value "honesty" was the best predictor of students' judgments of the morality of behavior in the work scenarios studied [9]. Thus, "honesty" is both a central virtue that guides our own behavior and a central value in judging the behavior of others.

In professional life, the values of engineers include both ethical values and others that are specific to the practice of their profession. As an example, possible values of a research engineer could be insightfulness, curiosity, integrity, and self-discipline. These would support the research engineer's career in many ways, resulting in what could be called "research excellence." It is good to remember the distinction that *values* are deeply rooted and often reflect what a person considers important in life/profession, while *virtues* are traits that are considered morally good and are cultivated over time.

Everyone Is Going to Win

That all sounds well considered and understandable, but how does it relate to the professional lives of engineers or engineering managers? We approach this relevant question through a brief vignette that tells a true story of a determined engineering student.

VIGNETTE Engineers Analyze, Design, Test, and Build

When Ovaska was an electrical engineering student, he wanted to learn everything about analyzing, designing, testing, and building prototypes or products. He regarded these subjects as "real things." Complementary courses, such as Engineering Ethics, Environmental Policy, and National Economy, did not interest him at all. Thus, he simply skipped those electives and did not understand why they even existed.

...

But decades passed, and now Ovaska is the co-author of this textbook, *Ethics for Engineers: Toward Ethical Behavior within Engineering Organizations*. So, what the heck happened to the "real engineer" over the years?

For an engineering student or a young professional, it might be difficult to see workplace ethics as an important *trigger* in a professional chain of consequences. But there exists a sort of chain reaction that ultimately brings something good to both the individual engineer and his organization—even to the shareholders who own the company. This abstract chain has the following consecutive links:

- Workplace ethics (practiced)
- Work atmosphere (enhanced)
- Personal wellbeing (getting better)
- Organizational effectiveness (improved)
- Benefits for shareholders ($++)

Thus in the long run, we can say that everyone is going to win. Of course, consequences are not the only reasons to behave ethically. And workplace ethics is not the only factor influencing the work atmosphere. But the important thing here is that the chain reaction actually exists. Once the existence of this progressive chain is *internalized*, engineering ethics is no longer seen as an isolated topic but as a useful—even natural—approach for both individuals and their organizations. One realizes that if an activity is undertaken by human beings, it *automatically* involves doing ethics. This is what happened to the "real engineer" in the vignette above, maybe after a decade in industrial research and development. It was a notable personal transformation. Think about it.

THERE ARE ALWAYS INDIVIDUALS BEHIND UNETHICAL CASES

The Diesel Emissions Scandal

When we read news about major unethical events—disasters, scandals, or "gates"—we usually first focus on their victims, and on the organizations, companies, even states seen as actors. And after this initial phase, we can begin to assess the more or less dramatic consequences for various parties, including customers, the companies themselves, the particular field of industry, and shareholders. A similar pattern of behavior also took place after the diesel emissions scandal of Volkswagen made headline news. Below is a brief overview of that scandal [10]. An extended discussion is presented later in this book (Chapter 8), which also examines a few other macro-level cases and, in particular, their short- and long-term consequences. These macro-level cases can be seen as cautionary examples.

In 2008, the German automobile manufacturer Volkswagen (VW) announced its strategy for the year 2018, in which the company expressed its objective to become the world's leading automotive manufacturer. It covered business from both an economic and an ecological point of view. The goal was to improve the fuel efficiency of their vehicles as well as to minimize harmful emissions. Unfortunately, it was done *partly* dishonestly. And in 2015, the U.S. Environmental Protection Agency (EPA) confirmed that Volkswagen had used specific software attached to the engine control unit, which hid the actual amount of nitrogen oxide emissions [11]. This was the beginning of the infamous diesel emissions scandal. The official EPA website, *Learn about Volkswagen Violations*, has the following description of that incident [11].

EPA: LEARN ABOUT VOLKSWAGEN VIOLATIONS

Volkswagen, Audi, and Porsche installed software on certain diesel vehicles that is designed to detect when the vehicle is undergoing emissions testing and turns full emissions controls on only during the test. The effectiveness of emissions control devices is reduced during all normal driving. This results in cars that meet emissions standards in the laboratory or testing station, but during normal operation, emit nitrogen oxides at levels up to 40 times the standard. This software is a "defeat device" that is prohibited under the Clean Air Act.

Since that came to light, numerous articles and reports have been published on various forums about this scandal. But the viewpoint of *individual actors* is largely missing. We do not know who the individual engineers and other professionals involved were, and there is no way to get the genuine explanations or motives for their respective actions. Nevertheless, some of the managers involved have been identified and prosecuted. In any case, this unethical behavior was carefully planned and carried out by a small insider arm of the huge Volkswagen organization. There was probably an informal chain of command from the manager in charge, who initiated the unethical process, to the software developers, who wrote the misleading code.

It is unlikely that the top management of the company directly started this emissions deception; the pressure may have been indirectly conveyed by setting overly high business goals for certain vehicles and certain market areas. And when the operative management found it impossible to achieve the goals purely with (a) innovative engineering methods, (b) novel solutions, and (c) top talent, they slipped into the miserable path of deception. And for a while, the outcome seemed to be successful. Sissela Bok wrote in her classic book: "If the incentives for achieving the goals—retaining one's job, most importantly, but also promotions, bonuses, or salary increases—are felt to be too compelling, the temptation to lie and to cheat can grow intolerable" [12] (p. 245). This appears to have been the case among some VW executives.

But what about the software developers who designed and implemented the fraudulent "patch?" It is easy to believe that their ultimate motive was nothing but altruistic—they were just serving their superiors and the company; whether it was Volkswagen directly or one of its trusted subcontractors. It is worth noting that while the software patch might be considered good from a pure technical standpoint, it was certainly not good on any other relevant standard. This reminds us of the difference between "good engineering" and "engineering that is good." On the other hand, the software developers who actually turned the deceptive scheme into a reality exhibited the virtues of determination, loyalty, respect, and self-discipline. But they did not act with fairness, honesty, integrity, and truthfulness. This is a profound ethical dilemma.

How many people were somehow involved in creating and implementing the fraudulent engine control software? Hard to say, but our guess is that the number must be in the low tens because the automotive industry has extensive testing and verification processes for embedded software and associated functions. So, a considerable number of VW professionals now bear the burden of having once been members of a small unethical arm of the company. Are they perhaps victims of their own unethical behavior?

Not Much for Learning Ethics

What could an engineering student or a young professional learn from this Volkswagen emissions scandal?—Well, not much except that it was bad and wrong, definitely unethical and illegal. Big incidents like this are usually so far removed from the life of an average engineer that it can be hard to relate to them. Hence, the pragmatic lessons behind macro-level cases can remain vague, as we usually do not get into their details—"they just happened somewhere."

Our approach in this book follows a simple lesson learned from an Ethics Advisory Council Workshop at Purdue University [13]: "It became clear that every engineering decision or project raises ethical issues and that engineering students need to understand this *day-to-day* aspect." (emphasis added)

Therefore, we decided to include in this textbook a representative set of micro-level case studies on *products* (Chapter 5), *employment* (Chapter 6), and *interaction*

(Chapter 7). And they provide a solid foundation for understanding and managing the ethical aspects of everyday engineering life, at the individual level. Such day-to-day vignettes are usually not available to the general public; they remain hidden in organizations and are not big enough to make news. These case studies provide a pedagogical platform for readers to recognize their own unethical intentions and make ethically sound decisions, in the future, when faced with similar temptations.

WHOSE BREAD I EAT, HIS SONG I SING

Binding Obligations of Employees

Let us still go back to the shocking diesel emissions scandal. How could it have been prevented? One thing is obvious: the employees at the bottom of the unethical command chain could not have stopped the deceptive plans—no matter how ethical their personal thoughts were. There was clearly an integrity problem at some level of the company's operative management, and it was spreading like an infectious disease. Whatever the ethical culture of the Volkswagen Group was in those years, the principles of ethical behavior were certainly not *internalized* at the top of that unethical chain of command; thus, cheating was acceptable because it could bring significant benefits and money. On the other hand, "Money must serve, not rule!" as Pope Francis stated in 2013 [14]. The risky deception happened because those in charge wanted it to happen *and* had a sense that they could distance themselves from any moral (or legal) responsibility. Under such circumstances, practically, nothing could have stopped it. But, of course, our reflection is just speculation.

In general, the concept of ethical behavior can be seen as a multi-layered and hierarchical entity for engineers working in a company (Table 1.2). Due to the obligations imposed by their employer, engineers *must* also consider the ethical standards of the company—even if these are less strict than their personal standards. And in possible conflict situations, engineers have a difficult decision to make—a decision that no code of ethics can make for them. Just as our own personal values need to be questioned from time to time, so do the values expressed by employers and shareholders.

TABLE 1.2
Hierarchy of Ethical Norms with Three Maturity Levels for Engineers to Consider

Norm Layer	Maturity Level 🙂	Maturity Level 😐	Maturity Level 🙁
(a) Company's ethical standards	Existing and communicated	Emerging	Unexpressed
(b) Individual's ethical standards	Thought out and internalized	Trained in (a), (c), and (d)	Extempore
(c) Professional code of ethics	Applied to day-to-day issues	Known	Not aware
(d) Moral theory	Advanced knowledge	Basic knowledge	Not aware

Money Can Buy Ethics

If we admit that the diesel emissions scandal was unstoppable, the next question would be: how can similar unethical incidents be prevented in the future? Our *optimistic* answer has two parts: (1) investors could push the companies they own to behave ethically; and (2) customers could demand ethical behavior from the companies they prefer. And, eventually, there would be company-pull, as well. But our *pessimistic* answer would be: when making profits and benefitting shareholders are seen as fundamental duties, then ethical considerations will always stand in danger of being ignored. Whether or not ethical behavior can dominate in contexts where responsibility is so easily diluted, we simply do not know.

Big shareholders have the control power, and they paid the painful bill for this emissions scandal, as well. In a month after its disclosure on 18 September 2015, when the U.S. EPA issued a notice of violation, the scandal had cost about 41% of the value of Volkswagen Group's ordinary share. On 17 September 2015—one day before the scandal became public—the closing price of the share was €167.80, and on 19 October 2015, it had fallen to €99.19 (*Source*: finance.yahoo.com).

Although the global decline in sales of Volkswagen vehicles was not that significant after the emissions scandal [10], it can be assumed that potential car buyers became more cautious when considering the purchase of a new car—regardless of brand. Therefore, the scandal affected that entire field of industry, too. Notably, in an advanced country like Finland where corporate corruption is somewhat rare, the VW emissions scandal did not receive as much attention as it did in the United States (where the use of defeat devices was illegal). In principle, customers of car manufacturers and other companies could have some power to initiate reforms toward an ethical corporate culture within the brands they prefer. But the power of customers is much less than that of big shareholders. In any case, it is practically impossible to initiate successful ethical reform in any large company at the lower levels of the organization, for example by individual engineers.

AN OPPORTUNITY FOR YOU

From the sections above, two motives can be identified *why* studying engineering ethics would be an opportunity for students and young professionals. One of those is related to ethical reforms in many workplaces [2]; these gradual reforms are demanded by the big shareholders and customers of large companies—thus "ethics is coming, be ready."

Let us think about this emerging trend through an interesting analogy with the environmental concerns of the past few decades. From the 1970s to the early 1990s, environmental concerns were merely a *nuisance* from the perspective of engineers and their companies—"those were just meant to disrupt business" was a common thought. Today, ethical concerns have a somewhat similar status in (too) many organizations. Moreover, environmental concerns (which, it must be noted, are grounded in ethical concerns) only came to the forefront when customers and investors, eventually politicians, started to pay attention to them. Although those issues always existed, many companies dodged their responsibilities, pretending that environmental matters were somebody else's concern. This has changed, and we believe the same will happen with ethical concerns more generally. Thus, the advanced ethical cultures of engineering firms will become *mainstream* only when their customers and investors start making such demands. From the point of view of environmental concerns, the most important motive today is the threatening increase in the concentration of carbon dioxide in the atmosphere; whereas in the ethical context, the key motivator would finally be the erosion of public *trust*. Think of the diesel emissions scandal from the point of view of lost trust; the customers and investors were impudently deceived. And a cynic would say that this was not the last such scandal.

The second motive behind studying engineering ethics is more personal. Engineers spend about eight hours a workday in their offices, meeting rooms, laboratories, and other workplaces. If the work atmosphere is not good, it can make employees anxious and less productive. In addition, bad feelings are easily transmitted to colleagues and even family members. Doing the right thing is important; acting against one's values ultimately leads to problems both at the organizational level and at the individual level. Thus, ethical behavior at the workplace—which leads to a better working atmosphere—increases personal wellbeing even after these eight working hours. This is the opportunity for you.

In addition to these two motives presented above, a third motive is emerging, and it is related to job retention in the long run. When organizations are downsizing and laying off some engineers (not masses), it is not only your present knowledge capital that maximizes the chances to retain your job. You are also expected to be a team player, at least a moderate communicator, and it helps if you are a committed and hard-working employee [12]. And this commitment actually has many faces, including a *commitment to ethical behavior* that is going to be a valued quality at some point in your career.

The reasons just presented in support of ethical behavior are decidedly practical. But there is a deeper reason for doing the right thing—who you are as a person is solely *your* responsibility. You might not be able to bring about ethical reform at the level of an entire company, let alone an entire profession. But at the end of the day, you do have the ability to determine what kind of person you are. Being current, collegial, and employed are certainly important but not at the cost of your moral character.

In Figure 1.1, we show the four elements of engineering ethics that form the core of this textbook. Moral theory is the solid foundation of good/bad and right/wrong; moral self-licensing is a convenient tool for explaining and understanding unethical actions; micro-level case studies highlight everyday issues; and macro-level cases provide a platform specifically for assessing consequences.

But if you are impatient and want a quick answer to the question of what you could do to become more ethical in your workplace, you could start applying an ethic of reciprocity known as the Golden Rule: "We must share the perspective of those affected by our choices, and ask how we would react if the lies we are contemplating were told to us" [13] (p. 93).

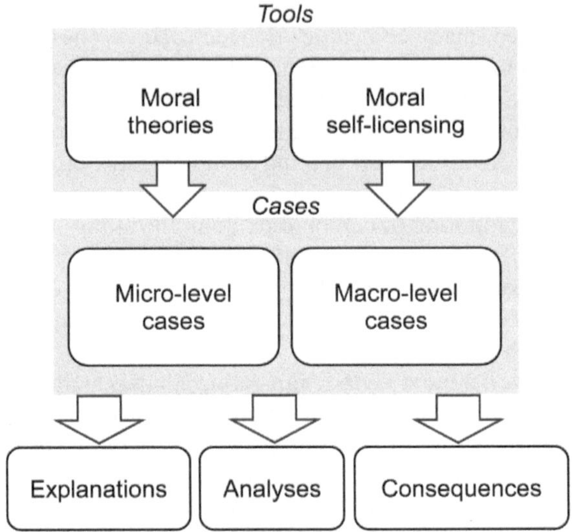

FIGURE 1.1 Principal elements for studying engineering ethics.

SUMMARY

Ethics for Engineers is a multidimensional topic that can be presented in many ways depending on the perspective chosen. Our holistic perspective is a combination of the perspectives of an engineer (Ovaska) and a philosopher (Brei), and throughout this text, we focus on the viewpoint of an individual actor. And to complement this, we present a timely company viewpoint in Appendix A written by the Senior Vice President, Ethics and Compliance of a major international company. Thus, all the necessary players are present.

We chose virtues as a framework for teaching ethics because it forms an intuitive approach to moral issues. And after presenting the concept of virtues, we found, interestingly, that the personal value "honesty"—one of the core virtues, and related to "integrity" and "truthfulness"—is also a good predictor of individuals' judgments of the morality of the behavior they observed in workplace scenarios.

Based on insider observations made in a few engineering organizations, there is a progressive chain between *workplace ethics*, *work atmosphere*, *personal wellbeing*, and *organizational effectiveness*. And when the existence of this progressive chain is internalized, ethics is no longer seen as an isolated subject but as a natural way of working for both engineers and their organizations.

The diesel emissions scandal of Volkswagen is a complex entity. It is practically impossible to get into the perspectives of its individual actors. However, it seems that under the special circumstances of those years of aggressive growth and hard competition, the deceptive plans were unstoppable. But the big shareholders would, indeed, have the power to make companies act ethically, in the future.

In conclusion, there is a movement toward ethical behavior in advanced engineering organizations. But every serious ethical reform can still be interpreted as a rare "trophy"; investors, customers, and employees are its ultimate winners.

Next, in Chapter 2, we show that unethical behavior, such as dishonesty, may come at an internal cost, for example, by increasing tension and harming the unethical actor's moral self-image. This familiar syndrome is called an *ethical hangover.* We also recognize and utilize an instructive analogy between the ethical hangover and the hangover from drinking too much alcohol.

CLASS EXERCISES

1. How would you define personal and professional ethics, and what are the principal differences between them?
2. How did your personal ethical standards develop into what they are today? And how do you distinguish good from bad and right from wrong?

   Class assignment: Compare the answers of individual students and create a common answer for the whole class.
3. What are the three most important virtues (see Table 1.1) in the working life of practicing engineers, and why are those of particular importance? Think of the three interrelated virtues of honesty, integrity, and truthfulness as a single meta-virtue.

   Class assignment: Compare the answers of individual students and create a collective answer for the class.
4. Create a list of values that you think are particularly important to a successful product development engineer and prioritize them. This list is likely to include both ethical values and others that are not related to ethics. Why exactly are these values important? You can use an AI chatbot (such as ChatGPT, Copilot, or Gemini) as your assistant.

   Class assignment: Compare the answers of individual students and create a common answer for the whole class. Six mutually agreed values would be good.
5. At the beginning of this chapter is a shortlist of actions (a–f) of typical fraud and misconduct in industrial organizations. Based on your personal values, rank these unethical actions in a descending order of severity.

   Class discussion: After all the answers are turned in, the instructor prepares a summary of the students' rankings. Discuss the summary and the fairness of the "voted order."
6. Due to the obligations imposed by their employer, engineers must also consider the ethical standards of the company—even if these are less strict than their personal standards. What could be the negative consequences if an engineer with high morals refuses to act unethically in his/her professional assignment, contrary to the instructions of his/her boss?
7. The text above has a section header "An Opportunity for You." It aims to highlight that studying engineering ethics would be an opportunity for

students and young professionals. But is it really a noteworthy opportunity for future engineers, or just another auxiliary topic?

Class debate: The instructor moderates a prepared debate between two volunteer teams. First, the class is divided into three teams; one of the teams is favoring the positive answer ("it is really a noteworthy opportunity"), the other is against it ("it is just another auxiliary topic"), and the third team evaluates the arguments presented. What are the objective conclusions of the evaluation team?

8. Find out what bribery is all about with the help of an AI chatbot. Is it always wrong to use bribery in working life? Justify your answer.
9. Create a mind map of this chapter.

REFERENCES

1. R. Shafer-Landau, *The Fundamentals of Ethics*, 6th Edition. New York, NY: Oxford University Press, 2023.
2. "Ethics Ambassadors: Promoting ethical culture on the front line." Stora Enso. Accessed: Jan. 19, 2024. [Online]. Available: https://www.storaenso.com/en/newsroom/news/2017/1/ethics-ambassadors-promoting-ethical-culture-on-the-front-line
3. D. Kim, B. K. Jesiek, C. B. Zoltowski, M. C. Loui, and A. O. Brightman, "An academic-industry partnership for preparing the next generation of ethical engineers for professional practice," *Advances in Engineering Education*, vol. 8, no. 3, 2020. Accessed: Apr. 1, 2024. [Online]. Available: https://advances.asee.org/wp-content/uploads/vol08/issue3/Papers/AEE-NAE-Brightman.pdf
4. "Integrity Survey 2013." KPMG Forensic. Accessed: Feb. 19, 2024. [Online]. Available: https://assets.kpmg.com/content/dam/kpmg/pdf/2013/08/Integrity-Survey-2013-O-201307.pdf
5. R. C. Solomon, "Corporate roles, personal virtues: An Aristotelean approach to business ethics," *Business Ethics Quarterly*, vol. 2, no. 3, pp. 317–339, 1992, doi: 10.2307/3857536
6. A. Brei, "Being loyal and being ethical," *IEEE Potentials*, vol. 41, no. 3, pp. 46–48, 2022, doi: 10.1109/MPOT.2020.2989712
7. Aristotle, *Nicomachean Ethics*, Book II, ch. 6, sec. 15. Princeton, NJ: Princeton University Press, 1984.
8. L. Sagiv, S. Roccas, J. Cieciuch, and S. H. Schwarz, "Personal values in human life," *Nature Human Behavior*, vol. 1, pp. 630–639, 2017, doi: 10.1038/s41562-017-0185-3
9. J. Finegan, "The impact of personal values on judgments of ethical behaviour in the workplace," *Journal of Business Ethics*, vol. 13, pp. 747–755, 1994, doi: 10.1007/BF00881335
10. I. Mačaitytė and G. Virbašiūtė, "Volkswagen emission scandal and corporate social responsibility—a case study," *Business Ethics and Leadership*, vol. 2, no. 1, pp. 6–13, 2018, doi: 10.21272/bel.2(1).6-13.2018
11. "Learn about Volkswagen Violations." EPA. Accessed: Jan. 30, 2024. [Online]. Available: https://www.epa.gov/vw/learn-about-volkswagen-violations
12. S. J. Ovaska, "Managing your career in a dynamic environment," *IEEE Potentials*, vol. 37, no. 3, pp. 24–26, 2018, doi: 10.1109/MPOT.2017.2764512
13. S. Bok, *Lying: Moral Choice in Public and Private Life*. New York, NY: Pantheon Books, 1978.
14. Pope Francis, *Evangelii Gaudium*, sec. 58. Accessed: May 28, 2024. [Online]. Available: https://www.vatican.va/content/dam/francesco/pdf/apost_exhortations/documents/papa-francesco_esortazione-ap_20131124_evangelii-gaudium_en.pdf

2 Ethical Hangovers

Learning Objectives

After studying Chapter 2, you will be able to

- Explain why the actors themselves can also be seen as victims of their own unethical behavior
- Take advantage of the commonsense model of unethical behavior to control moral licensing
- See the instructive analogy between an ethical hangover and an alcohol hangover
- Understand unethical organizational culture as an active source of spreading unethical behavior

It is useful to be aware that sometimes even the actors themselves can be seen as *victims* of their own unethical behavior, such as dishonesty [1] (p. 24). This can be a new perspective for anyone—whether they engage in occasional dishonesty in their personal or professional lives. Here, our focus is on the day-to-day professional lives of engineers, but much of this discussion is relevant in other contexts, as well.

Related to this, Zhong and Robinson recently addressed an important question that has not received much research attention so far: "How does negative behavior affect actor's emotions, needs, thoughts, and organizational success" [2]? They made a tentative conclusion in their review article on Actor-Centric Outcomes of Negative Behavior: "Engaging in negative acts is a two-edged sword for actors and its costs seem to slightly prevail over its benefits" [2]. This kind of research on the outcomes of actors' negative behavior can help to better understand the reasons why negative behavior occurs and persists in organizations. Unethical behavior, the underlying theme of this chapter, covers much of what they use the term *negative behavior* for. So, while it is obvious that "bad actors" benefit from their negative behavior—why else would they act so, if not for personal gain?—it is clear that this sort of behavior involves costs. And it is one of these costs that we now turn to.

CONSEQUENCES FOR ACTORS THEMSELVES

Generally speaking, whenever we are victims of something, we usually suffer in some way. This is true, of course, when we are on the receiving end of someone else's dishonest behavior. But acting dishonestly also involves a kind of suffering for the actor because, for example, it increases tension and damages the dishonest actor's moral self-image [3]. We call this common disorder an *ethical hangover*, the symptoms of which can last for a moderate or longer period of time—even years. A

DOI: 10.1201/9781003485520-2

TABLE 2.1
Typical Symptoms of Ethical Hangovers

Symptom
Damaged moral self-image
Lowered self-esteem
Cognitive dissonance
Self-accusations
Repetitive thoughts around the unethical action
Feeling guilty
Feeling ashamed
Anxiety in different forms
Nightmares
Sleeplessness
Internal tension in general
Tensions with the people involved
Tension in related places
Tension during similar tasks
Decrease in work efficiency

representative set of possible symptoms is listed in Table 2.1. And any ethical hangover can consist of some combination of these (and other) symptoms.

But why would anyone inflict such inner suffering upon themselves? Sissela Bok gives four reasons for lying and other forms of dishonest behavior in her book, *Lying: Moral Choice in Public and Private Life* [1] (pp. 78–86):

- Avoiding harm
- Producing benefits
- Fairness (e.g., giving people what they "deserve")
- Veracity (e.g., lying to protect the truth)

The first two of these are the most commonly used motives behind dishonesty, and they often go hand in hand. And when the unethical *temptation* becomes strong enough, the actor is willing to accept the ethical hangover that may follow. In this case, the actor's self-discipline has collapsed. These temptations are primarily due to desires and inclinations. Interestingly, though, not everyone gets an ethical hangover after a particular act of dishonesty, and the intensity of the ethical hangover can vary from person to person. Fortunately, it is possible to resist temptations if the self-discipline is strong enough.

Commonsense Model of Unethical Actions

In order to comprehend the step-by-step nature of unethical actions, we propose a straightforward three-step model of unethical behavior in Figure 2.1. The model proceeds along these lines: Step 1: After an unethical temptation appears, *moral*

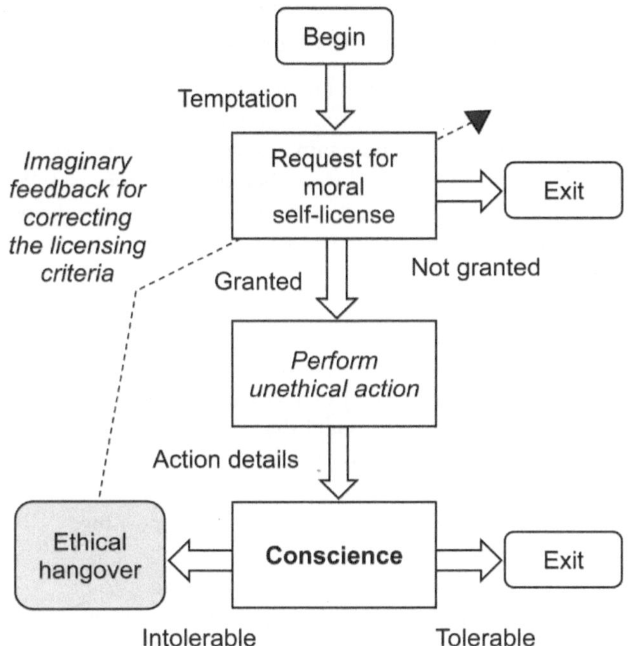

FIGURE 2.1 Commonsense model of unethical behavior.

self-licensing is needed before the corresponding action is allowed to oneself. This involves giving oneself permission (license) to perform a bad action by considering it in comparison to a good action that one has already performed or expects to perform in the future [4]. If a satisfactory license cannot be formed, this temptation is effectively over. But if the individual in question grants the license to himself, he allows himself to perform that particular unethical action. Step 2: When the action is performed, all related details are delivered to the actor's *conscience* for processing and analysis [5]. Step 3: If the conscience classifies the action as "tolerable," the event is over. However, if the unethical action performed is classified as "intolerable," an ethical hangover will inevitably result. The dashed feedback connection in this commonsense model is explained below in the section "Every Ethical Hangover Is a Chance."

Two points bear mention here. First, "tolerable" is not the same as "right." An actor whose conscience has labeled an unethical action "tolerable" has knowingly done wrong but has rationalized the bad behavior. Additionally, it should be emphasized that the two decision blocks in this model are actor-specific because everyone is an individual with their personal criteria for moral licensing as well as for moral judgments. In Chapter 4, we review the central issue of moral self-licensing in detail, along with other explanatory tools in the field of behavioral ethics.

Next, we go through a straightforward example illustrating how the proposed model of unethical behavior works in practice. The underlying case is presented in the following fictional vignette.

VIGNETTE An Engineering Manager's Mistake

Kurt was an accomplished engineering manager who lived and worked in Northern Finland, just below the Arctic Circle. His company had its annual Management Convocation at a mountain resort hotel about 300 km north of Kurt's home. The summer meeting was scheduled to begin at 9:00 a.m.—so, Kurt had to wake up very early because it would take about four hours by car from his house to the meeting venue. But Kurt slept a little too late and thus he was in a hurry.

At the very beginning of his trip, there was 10 km of road construction on the highway; the new asphalt surface was already there, but the road markings were still missing and thus the speed limit was only 50 km/h. Kurt was impatient and felt a strong temptation to overspeed—maybe up to 100 km/h—but not faster because the road was quite winding and hilly.

His moral self-license was based on the following elements:

- There was practically no traffic this early in the morning.
- The road surface was perfect.
- There was no automatic speed detection system in use.
- The weather was excellent, as it usually is in the late June.
- He was an experienced driver with no previous accidents.
- His car was new and safe.
- He was in a hurry to attend the meeting where he was to receive the Corporate Innovation Award.

After a short (but subjective) thought, he smiled to himself and accelerated his car to 100 km/h; it would not harm anyone. Kurt had now licensed himself to perform not only an unethical but also an illegal act.

After a while, there was a rather steep uphill, and behind the hill top, a herd of about 20–30 reindeer was crossing the highway. It came as an unexpected shock to Kurt; he pressed the brake pedal hard but could not stop his car in front of the herd of reindeer. Very probably, he could have done it easily if his speed had been 50 km/h (the braking distance is directly proportional to the square of the car's speed).

It was a bloody crash: Kurt was injured himself and had to spend a couple of weeks in the hospital, his car was badly damaged, and about 10 reindeer were killed immediately or seriously injured (had to be put to death later). These and other action details were "delivered" via reflection to his conscience for processing and analysis.

In the hospital, where Kurt had plenty of time to think, his conscience classified that unethical and illegal action as "intolerable." Because of this, Kurt suffered a severe and long-term ethical hangover; he was a victim of his own unethical behavior. And all this suffering for the sake of saving six minutes, in the best possible case—the benefit that did not materialize.

So, what is the lesson from this example?—Well, whenever we construct a moral license for ourselves, it is practically impossible to consider all relevant scenarios, and all too easy to overlook their probabilities. Such is also happening to a number of engineers every day, in their professional life. As we know, the management of uncertainty is challenging.

CONSCIENCE AS A JUDGE

Conscience is a kind of "mystical" meta-function in our minds; we probably heard about it already early in our childhood. And we learned that we can have a bad conscience if we do something that is wrong. But it is likely that most people have not spent much time thinking *what* the conscience actually is. A brief discussion is therefore useful, because conscience is the key to triggering an ethical hangover, as shown in Figure 2.1.

The Pocket Oxford Dictionary defines *conscience* as "moral sense of right and wrong (good or clear, bad or guilty)." Here, the imprecise word "sense" opens the door to individual-specific outcomes in particular cases. But that is about all what we can say based on this definition.

More helpfully, Giubilini defines the purpose of conscience in the *Stanford Encyclopedia of Philosophy*: "Through our individual conscience, we become aware of our deeply held moral principles, we are motivated to act upon them, and we assess our character, our behavior and ultimately our self against those principles" [5]. And then he continues with something more definitional: "conscience is defined by its inward looking and subjective character, in the following sense: conscience is always knowledge of ourselves, or awareness of moral principles we have committed to, or assessment of ourselves, or motivation to act that comes from within us (as opposed to external impositions)." There is a lot contained in these few lines, and they would take some effort to unpack.

For our purposes, it will suffice to understand that conscience is a kind of knowledge, of both factual and evaluative sorts. The "moral principles we have committed to" corresponds to the "Individual's ethical standards" in Table 1.2. But unfortunately, these can be overridden by "Company's ethical standards," which may be less stringent than one's own standards and can, in certain situations, have a higher priority. This is the reason why an engineer's conscience may sometimes classify his company-driven actions as "intolerable," even if he himself had wanted to behave ethically. Figure 2.2 shows an imaginary example of how a relatively large number of moral self-license requests ends up in a fairly small number of intolerable actions that lead to an ethical hangover.

An important distinction is worth noticing here. Moral self-licensing, as we use it in our model, is a voluntary act. It is something the actor does intentionally. But the evaluative job of the conscience is, by contrast, involuntary. It happens unintentionally, whether we would like it to or not. As long as a person has licensed and performed a bad action (Steps 1 and 2 in our model), that person's conscience will spontaneously render a verdict. Thus, we have explicit control over the decision-making phase in Figure 2.1 but no explicit control over the subsequent judgment

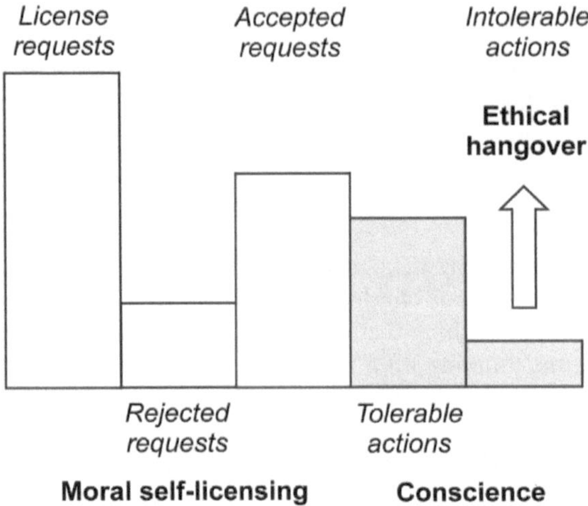

FIGURE 2.2 From license requests to intolerable actions that lead to an ethical hangover (example).

phase. This seems analogous to the distinction between legislative and judicial branches of some modern governments. The legislative branch creates laws, setting the framework for what is deemed acceptable or unacceptable, similar to how individuals craft criteria for moral self-licensing. And the judicial branch interprets and applies these laws to specific situations, ensuring justice is served. Correspondingly, the independent conscience reflects on and evaluates one's actions, leading to ethical hangovers when actions do not align with one's moral standards.

Although the idea of a conscience will be familiar to nearly all of us, there is no universally accepted definition of the concept itself. It has been interpreted in different ways throughout history [5], each reflective of a particular cultural, religious, or normative milieu. But since this textbook is intended for engineering students and young professionals, perhaps a pragmatic definition of conscience could be expressed in terms more familiar to computer engineers:

> Conscience is an intelligent processing unit operating in the background, which collects specific and detailed information about the individual's behavior; it is parameterized by learning from past experiences, and it adapts its parameters to changes in the action environment; the parameters of personal and professional life differ slightly from each other; it monitors an individual's behavior and alerts when the boundaries from right to wrong or good to bad are crossed; such an alert activates the ethical hangover.

Remaining in a technical context, we might also make the following analogy: automotive engineers have designed several warnings and alarms into our cars, such as check engine, coolant temperature, oil pressure, and brake system. And we are used to acting when one of these dashboard lights goes on; it is not wise to ignore such alerts. How about applying the same practice to possible "conscience alerts," too?

ANALOGY WITH ALCOHOL HANGOVERS

The U.S. National Institute on Alcohol Abuse and Alcoholism (NIAAA) tells us the following: "A hangover refers to a set of symptoms that occur as a consequence of drinking too much. Typical symptoms include fatigue, weakness, thirst, headache, muscle aches, nausea, stomach pain, vertigo, sensitivity to light and sound, anxiety, irritability, sweating, and increased blood pressure" [6]. And it concludes, "ultimately, the only surefire remedy for a hangover is to avoid getting one by drinking in moderation or choosing not to drink." This is sound advice, grounded firmly on the biological and chemical sciences.

Interestingly, we can recognize an analogy between an ethical hangover and an alcohol hangover. An ethical hangover refers to a set of symptoms (including guilt, anxiety, and tension) that occur as a consequence of licensing and performing unethical actions. Avoiding unethical behavior is the best way of avoiding an ethical hangover, and this would be the advice of the moral philosopher. Another way of avoiding this sort of hangover would involve the conscience classifying an unethical action as "tolerable." Understanding how that is possible is the domain of the social scientist.

Getting and Suffering

Hangovers of either type—alcoholic or ethical—are self-inflicted. This is illustrated by the following fictional vignette, where we present an alcohol case followed by an ethics case with clear parallels.

VIGNETTE So Near and Yet So Far

ALCOHOL CASE

Erich was an electrical engineer who was going to have a couple of beers with his old college buddies. They met once or twice a year. This time, Walter had chosen the tavern, and two other buddies were also there. But Erich was surprised that Egon was not around. When he asked why, Walter laughed and said, "Egon is a sheep who only drinks soda pop and complains about everything. But the rest of us know how to have a good time!"

Nevertheless, Erich's intention was to drink only in moderation. But after a couple of beers, they took a couple of shots, and then more and even more, until they were all heavily drunk. At first, they were having fun, but later they were just goofing around. Finally, they quietly departed in different directions.

The next morning Erich felt terrible, he had all possible symptoms of an alcohol hangover. Of course, he could not go to work, but he had to stay in bed until the late afternoon. There he had time to think—why did this happen to him again? He was only supposed to have a couple of beers with those guys.

ETHICS CASE

Erich, who was the head of a company's product development department, attended a meeting where they discussed the launch of their new process automation system for the South American market. The meeting had been called by Walter, the vice president of marketing. Also, the vice president of technology and the sales director were present. But Erich was surprised that Egon was not there because his department was in charge of maintenance tools and related procedures. When Erich asked the reason for that, Walter arrogantly replied, "Egon is a damn sheep who always asks questions ('questioning is disloyalty') and ruins the whole atmosphere. But we know how to make wise decisions."

Walter proposed that they would soon launch the new product in the South American market because according to the company's strategy, they definitely wanted to enter that growing market. Their sales and maintenance organizations were just starting in Brazil.

Erich was worried. All the product's environmental tests (temperature, humidity, vibration, and electromagnetic compatibility) were still missing, no field tests had been done yet, and the user interfaces had not been tailored to the language/cultural area in question. Furthermore, their service network in Brazil was rudimentary, with local technicians only having basic training for a few legacy products. The risk would be high and multifaceted.

Walter and the others ignored Erich's objective concerns, and Erich decided to be quiet because he did not want to make Walter angry—maybe someday, Erich would also be a vice president. However, he understood very well that it would be his department that would bear the burden of all the possible trouble.

After some discussion and preliminary planning, they made a "unanimous" decision to (prematurely) introduce the new product to the South American market, but starting in Brazil. When the meeting was over, Erich walked toward his office in a pensive mood, feeling bad that he had again agreed to something that went against his core values. He had an ethical hangover.

So, what is the common thread between these two cases? The first similarity is that they are both group cases—and the group's role is crucial. Sissela Bok writes in her book: "The shared belief that 'we are a wise and good group' inclines them to use group concurrency as a major criterion to judge the morality" [1] (p. 97). And thus, "any means we decide to use must be good." This kind of licensing assumption is present in both cases. Also, the drinking/decision-making went clearly beyond Erich's initial thinking. And eventually, he was alone with his alcohol/ethical hangover. The night of drinking and the high-risk "unanimous" decision led to the embarrassing consequences of an alcohol hangover and ethical hangover, respectively.

Common Myths about Hangovers

Although many remedies for alleviating alcohol hangovers are mentioned on the web and particularly in social media, none has been scientifically proven to be effective [6]. The same is true for ethical hangovers. Below is one common myth and an equivalent fact for each.

Myth A: Certain actions, such as drinking coffee or taking a shower, can prevent or cure an alcohol hangover.

Fact A: There is no cure for an alcohol hangover other than time [6].

Myth E: The ethical hangover can be relieved if its burden is shared, for example, with one's boss [7].

Fact E: There is no cure or remedy for an ethical hangover other than time, but a much longer time than with alcohol hangovers.

Every Ethical Hangover Is a Chance

On the other hand, could ethical hangovers have any positive consequences? Yes, maybe. We can see our judgmental conscience also as a kind of *motivator* to behave (more) ethically. Because after the negative experience of an ethical hangover, we are concretely reminded that we did something bad or wrong, and it can be an activator to avoid such unethical behavior the next time we are similarly tempted. If that is the case, we have really learned something.

This learning opportunity is illustrated in the model of Figure 2.1 with a dashed line as feedback from the ethical hangover block to the moral licensing block. Such imaginary feedback is used to correct the (too loose) moral licensing criteria—similar what is done in control engineering applications. And here, the ultimate goal could be to avoid ethical hangovers completely, or at least to avoid severe ethical hangovers, for instance. But the ability and willingness of individuals to make such corrections varies considerably [1] (p. 243).

If an actor is suffering from an ethical hangover but is not determined enough to change his behavior on his own, help is available. Fortunately, engineering ethics can be practiced and learned either individually or in classes and seminars. Flexible online courses could be best suited for working engineers. And a good starting point would be Sissela Bok's eye-opening book, *Lying: Moral Choice in Public and Private Life* [1]. Although written by an accomplished philosopher, it is also suitable for motivated engineers looking for help with their ethically questionable behavior. Crucially, however, none of these can help a person who does not really want to be helped. Here, again, is another similarity between chronic immorality and chronic alcohol abuse.

There are also actors who suffer from ethical hangovers but do nothing to get rid of them or change their behavior to be more ethical. These people only increase their ethical-hangover load, which leads to deterioration in personal wellbeing. This is a somewhat similar situation to those who have alcohol hangovers over and over again but do nothing about their problem. A positive first step in a situation like this would be to admit that "I've got a problem."

But what about the cases where the actor did not get an ethical hangover after his unethical behavior? His conscience classified the action taken as "tolerable," and he did not get the *chance* that an ethical hangover would have offered. He does not feel that he did anything wrong, as it was a strongly self-licensed action. In such cases, his organization's possible ethics program may still prove useful. If the program is pedagogically stimulating and the management is genuinely committed to it, then even an unethical actor, who does not usually suffer from ethical hangovers, could gradually transform his behavior to be more ethical—"I wanna do what's right." Unfortunately, this is not the outcome for everyone.

We hope that the commonsense model of performing unethical actions described here might help practitioners to gain more insight into the actual process of getting an ethical hangover, and thereby learn to control and manage their tolerance for unethical temptations. Engineers know that it is practically impossible to control any process or system satisfactorily if we do not understand its details. So, every piece of additional information is helpful.

UNETHICAL ORGANIZATIONAL CULTURE

An engineering organization consists of a group of collaborating engineers and other professionals with a particular common purpose, such as product development. The size of the organization may vary considerably. But if we talk about an *unethical organizational culture*, what does it mean? Basically, managers create and maintain the culture within their organization and pass it on to their subordinates. Even a single unethical individual at the top of the management chain can be the infectious source of unethical actions because they lead by their (negative) example. Therefore, an organization's ethical culture is heavily influenced by the ethical intelligence of its CEO. Upon the departure of a CEO and the arrival of a new one, it can take just a few weeks for this culture to be significantly altered.

To the extent that an organization's ethical culture is believed to depend on the benefits it offers, much hinges on the stability of profits. It can even be argued that if an empirical study were to show that the link between ethics and profit is tenuous or non-existent, the (economic) incentive to foster an ethical culture would vanish. Fortunately, other incentives, such as compliance with regulations and maintaining a positive reputation, could still support ethical business practices—or "doing what's right"—within the organization.

Cross lists seven organizational qualities that promote and often lead to unethical behavior in business, including engineering [8]; they are given below:

1. Conceit
2. Cronyism
3. Cult
4. Dread
5. Desperation
6. Disregard
7. Disdain

To encourage active learning, we provide this plain list here as a starting point for exploration. In an upcoming Class Exercise, students will search for the definitions and implications of these qualities using Cross' original, open-access publication [8].

After introducing those specific qualities, Cross concludes: "If many of these qualities are present, however, that—itself—does not mean that unethical conduct will result. But there is a high likelihood that it will." And managers carry a heavy responsibility in this regard. They need to ask themselves a few critical questions to ensure that every decision and action is ethical, as discussed in the pragmatic article by Hyman et al. [9]—which contains a checklist of questions designed to enhance one's ethical sensibility.

Problems with an organization's ethical culture can sometimes lead to *macro-level* unethical incidents, such as the Volkswagen diesel emissions scandal outlined in Chapter 1. Or they can lead to diverse *micro-level* cases that can incrementally worsen the working atmosphere, and thus reduce the effectiveness of the organization and even increase employee turnover. When trust in the company's management decreases, the commitment of employees also decreases. Trust can only flourish on the basis of respect for the truth. At some point, ethical problems within any organization become visible from outside the company, as well.

Like a Contagious Disease

Possible unethical behavior of managers is easily transmitted to their closest subordinates, and further to their colleagues, and so on. And if an individual's "immunity" is not sufficient, then he will likely succumb to the unethical pressure. The spread of unethical behavior in an organization may appear like a local epidemic. In the "Ethics Case" above, the unethical spirit of the leadership team was transmitted to Erich; thus, he acted against his own values. Among the virtues violated, there were fairness, respect, and responsibility. And from Erich's point of view, there was also a tricky conflict: easy risk-taking versus heavy risk-bearing. As a member of the decision-making team, he took the risk, and as the head of his department, he would ultimately bear its likely consequences.

Furthermore, accepting and adopting certain unethical behavior of managers can lead to a "quiet invitation" to join a group of insiders, which can, among other things, open doors to career advancement and other benefits. This is called "cronyism," as noted in reference [8]. Unfortunately, this is still happening today, and some engineers see it as a positive opportunity. However, it is merely another potential source of ethical hangovers.

Toward Ethical Reforms

More than three decades ago, Phillip B. Crosby stated: "Quality rests in the hands of management, not in the quality control department" [10]. Although this is widely accepted today, back then such an opinion was somewhat revolutionary. And it was the new beginning of quality reforms in many companies worldwide.

Currently, in some companies with emerging ethics cultures, the management of ethical issues has been *delegated* to ethics committees or part-time ethics coordinators. However, this encourages the idea that ethics is the concern of a select few whose job it is to care about such matters. The problem with this is that the ethical culture of the organization as a whole may not improve, even if the company's code of ethics (or code of conduct) exists and has been communicated to all employees [1] (p. 246). Therefore, the same principle that Crosby proposed for quality would also be natural for ethics. Bearing in mind *both* the fact that management has a responsibility to create a culture where ethical behavior is the norm *and* the fact that, ultimately, our moral characters are our own responsibilities, we might amend Crosby's statement, thus creating an ethical culture is in the hands of management; fostering that culture is in everybody's own hands.

In the years to come, after profound cultural reforms and dedicated exemplary leadership, future engineers will hopefully suffer fewer ethical hangovers, and the world will see fewer unethical scandals. As Crosby once claimed that a systematic drive for quality will pay for itself [9], could it be that a similar systematic drive for ethics would pay for itself, too? Yes, we believe that value-driven ethics programs would pay for themselves in the workplace. The reforms of ethical cultures in organizations are discussed further in Chapter 10 and Appendix A.

SUMMARY

Why should individual engineers behave ethically if there is pressure to do "whatever it takes" to achieve business goals? That is a significant question to ponder—both personally and professionally. Hopefully, this chapter provided a new way of thinking about it. Our unusual perspective is related to the fact that sometimes the actors are victims of their own unethical behavior. And who wants to be a victim of anything?

In practice, engineers are bound by several different standards. Some of these may be referred to as "industry standards," and often relate to technical matters. Others may be described as "company standards," and relate to, among other things, the expectations placed on a professional by an employer. These standards can be explicit or implicit, and are reflected in mission statements, codes of conduct, and company culture. In addition, there is what we may call the "moral standard." A company can go some way toward establishing this for its employees, but ultimately the responsibility of identifying and adhering to this standard is on each individual person. Because employees have obligations to themselves, their employers, their profession, their communities, and beyond, the potential for conflicting standards is real. And when company standards are less stringent than an individual's moral standards, ethical hangovers can result. To address this, management ought to work to bring about positive cultural reforms in companies as a way of discouraging unethical behavior and thus preventing ethical hangovers. At the same time, professionals need to take responsibility for their actions and do what they can to resist the temptation to do wrong.

To close this chapter, we introduce the concept of "moral conversion." Sometimes, getting caught can trigger a conversion experience (as "every ethical hangover is a chance"). One may gradually learn to suppress one's conscience through repeated instances of wrongdoing without consequences. However, it is natural human behavior to keep pushing boundaries to see how far one can go. When one goes too far, resulting in catastrophic consequences, a complete reversal can occur. This is the basis for prison reform. Wouldn't it be much better to initiate one's own moral conversion rather than wait for a total collapse to force reform?

Next, in Chapter 3, we present different moral theories and one professional code of ethics with a moral theoretical basis. These will help us as we reflect on our own moral standards, and will provide the necessary tools for analyzing and assessing the micro-level behavior of engineers in their everyday tasks.

CLASS EXERCISES

1. How were the virtues *fairness*, *respect*, and *responsibility* violated in the "Ethics Case" of the vignette "So Near and Yet So Far?"
2. Look for a reported case of dishonest behavior in either professional or public life, and create a speculative moral license for it (see Table 2.1). Use the license in the vignette "An Engineering Manager's Mistake" as a model.

   Class discussion: Compare selected dishonest cases and students' speculative moral licenses. How compelling are those licenses? Remember that these unethical actions were performed in reality.
3. Form an acting group of four volunteer students. The group first prepares a script for a role-play based on the "Ethics Case" from the vignette "So Near and Yet So Far." And then they perform it in front of the class.

   Class discussion: What new issues, if any, related to ethics in engineering did the role-play raise compared to the text of the book alone? The actors' viewpoint versus the audience's viewpoint.
4. Choose one of the qualities (1–7) of unethical organizations, and examine its rationale in Cross' original article [8]. Alternatively, the instructor could allocate those seven qualities to students. Summarize the rationale in no more than three bullet points.

   Class discussion: Are there actual reasons or possible ways to get rid of such corrupting qualities?
5. What is meant by the concept "conscience?" Form your condensed but easy-to-understand answer from Giubilini's passages quoted in the section "Conscience as a Judge." You can also check out his original article for more information [5].
6. Use one of the AI chatbots, like ChatGPT, Copilot or Gemini, and find a thorough definition for the term "conscience."

   Class discussion: How do the responses of different chatbots differ from each other—are there any significant differences?
7. There is an analogy between ethical and alcohol hangovers. Could the methods utilized in overcoming alcoholism be adapted and applied to unethical behavior? Outline a procedure for recovering from chronic unethical behavior.

   This assignment is suitable for groups of three students.
8. Does the proposed analogy between an ethical hangover and an alcohol hangover offer any instructive benefit or added value? Consider both professional and personal viewpoints.

   Class debate: The instructor moderates a prepared debate over Question 8. First, the class is divided into three teams; one of the teams defends the positive answer ("yes"), one team supports the negative answer ("no"), and the third team evaluates the arguments presented. What are the objective conclusions of the evaluation team?

9. How could the ethical climate be measured in an engineering organization? Without this kind of measurement, we cannot know how ethical or unethical the organization is. Use Google Scholar to search for relevant literature.
   Class discussion: Are there such objective measures for the ethical climate that would be practical to implement in an engineering organization?
10. Create a mind map of this chapter.

REFERENCES

1. S. Bok, *Lying: Moral Choice in Public and Private Life*. New York, NY: Pantheon Books, 1978.
2. R. Zhong and S. L. Robinson, "What happens to bad actors in organizations? A review of actor-centric outcomes of negative behavior," *Journal of Management*, vol. 47, no. 6, pp. 1430–1467, 2021, doi: 10.1177/0149206320976808
3. O. Weisel and S. Shalvi, "Moral currencies: Explaining corrupt collaboration," *Current Opinion in Psychology*, vol. 44, pp. 270–274, Apr. 2022, doi: 10.1016/j.copsyc.2021.08.034
4. D. A. Effron, "Beyond 'being good frees us to be bad:' Moral self-licensing and the fabrication of moral credentials," in *Cheating, Corruption, and Concealment: Roots of Unethical Behavior*, J.-W. van Prooijen and P. A. M. van Lange, Eds., Cambridge, UK: Cambridge University Press, 2016, ch. 3, pp. 33–54, doi: 10.1017/CBO9781316225608
5. A. Giubilini (2023), "Conscience," in *The Stanford Encyclopedia of Philosophy*, E. N. Zalta and U. Nodelman, Eds. Accessed: Jan. 26, 2024. [Online]. Available: https://plato.stanford.edu/archives/fall2023/entries/conscience/
6. "Alcohol's effects on health: Research-based information on drinking and its impact." NIAAA. Accessed: Jan. 26, 2024. [Online]. Available: https://www.niaaa.nih.gov/publications/brochures-and-fact-sheets/hangovers
7. A. Brei and S. J. Ovaska, "An ethical hangover: A young professional's case," *IEEE Potentials*, vol. 43, no. 3, pp. 6–8, 2024, doi: 10.1109/MPOT.2023.3342431
8. J. Cross, "The seven deadly sins of unethical organizations," *Ethikos*, vol. 28, no. 4, pp. 3–6, July/Aug. 2014. Accessed: Jan. 26, 2024. [Online]. Available: https://assets.corporatecompliance.org/Portals/1/PDF/Resources/ethikos/scce-2014-07-ethikos.pdf
9. M. R. Hyman, R. Skipper, and R. Tansey, "Ethical codes are not enough," *Business Horizons*, vol. 33, no. 2, pp. 15–22, 1990, doi: 10.1016/0007-6813(90)90004-U
10. P. B. Crosby, *Quality Is Still Free: Making Quality Certain in Uncertain Times*. New York, NY: McGraw Hill, 1996.

3 Doing Ethics
Theory and Code

Learning Objectives

After studying Chapter 3, you will be able to

- Understand key terminology in the field of ethics
- See moral theories as versatile tools for practitioners
- Apply moral theories to evaluate ethical issues and justify moral judgments
- Use the IEEE Code of Ethics to evaluate ethical issues and justify professional behavior

Most of us would say that we have got a pretty clear and well-developed sense of right and wrong. Often, all that means is that we have figured out how to fit in among the people with whom we live and work. And once we know how to "get by" in those contexts, we usually think no more about it. Understanding morality in this way runs roughly along the lines of a view called Moral Relativism [1]. According to the most common version of this view, moral right and wrong are simply a function of what a particular culture accepts or rejects. If an action falls under some custom or convention that is deemed acceptable by a group of people, then that action will be called right. Accordingly, an action will be called wrong if it falls outside of cultural norms. For a *cultural relativist*, then, being moral amounts to no more than doing whatever the other people in your culture do.

But look, if you had been brought up by a society that permitted cannibalism, you would likely have a rather favorable view of that practice. Similarly, if you had been brought up in a place and time where women were not allowed to vote, own property, pursue careers, or even speak in public, you would probably have developed a fairly dim view of the abilities and value of women. Clearly, what a society happens to accept is no reliable indication of what is morally right. You would not need to think for very long before you came up with examples of traditions and practices that were accepted by societies of the past (or perhaps the present) but that are demonstrably wrong. Female infanticide, human trafficking, foot binding, and slavery come to mind.

No—if we *truly* want to know about the moral status of cannibalism and subjugating women, we are going to need to look deeper than what a society accepts and practices. We are going to need to understand why and how something could be right or wrong, good or bad. We are going to need some moral theories.

DOI: 10.1201/9781003485520-3

MORAL THEORY

When we talk about theories of morality, we are talking about different accounts of how to live well. Moral theories offer explanations for why actions are right or wrong (or why people are good or bad). Usually, these explanations can be condensed into *moral principles*, which are statements that provide some guidance when we think about how to live and act—something like, "Actions are right if they…" or "Only act in ways that…." One common misconception about moral theories and principles is that they *make* actions right or wrong. Not so! Moral theories only help us make sense of the factors that contribute to something's *moral status* (its rightness or wrongness, goodness or badness) and provide guidance for various different situations in life. A similar point might be made of the law (that is, the legal standard articulated by municipal, national, and federal codes and regulations). Murder is illegal as well as wrong. But murder is not wrong *because* it is illegal. Rather, it is the other way around—murder is illegal because it is wrong. In a similar way, murder is condemned by every moral theory. But murder is not wrong *because* it is condemned by moral theory. Rather, it is condemned by moral theory because it is wrong. Think about this fundamental difference.

In addition, the reasons offered for murder's wrongness will vary, depending on the moral theory one considers. Many different moral theories have been articulated by many different moral philosophers. We will focus on three that have dominated *ethics* (that is, the philosophical study of morality) in the so-called Western tradition:

1. Utilitarianism
2. Deontology
3. Virtue Theory

Of course, these are not the only moral theories. And there are different interpretations and variants of each of them. To keep things manageable and understandable, we will present these three views in their most classical formulations, as expressed by their most prominent defenders. As we will see, they offer some interesting and worthwhile perspectives on the factors relevant to determining right and wrong (or good and bad).

ACTIONS VS. CHARACTER

One more general point should be made before we turn to the specific moral theories. There are two basic approaches that one might take when doing ethics. One is to focus on *actions* and to ask the question "What should I do?" The other is to focus on *character*, where the main question is "Who should I be?" The moral theories that focus on character tend to emphasize the importance of *human nature* (that is, the qualities that are shared by all human beings) and what it means to live a good human life. On the other hand, the moral theories that focus on actions recognize the fact that the things we do are preceded by motives and followed by consequences. That fact could be illustrated by the following process:

MOTIVES ⇒ ACTIONS ⇒ CONSEQUENCES

As we will describe below, some action-focused theories emphasize the importance of motives, while others pay more attention to consequences. With that in mind, we turn to some theories.

THE UTILITARIAN PERSPECTIVE

The first moral theory we will discuss is Utilitarianism [2]. The name might suggest that this is a view which equates rightness with usefulness. After all, *utility* can mean *useful* or *convenient*—as in the case of a sports utility vehicle or a utility belt. However, this is not the sense of utility that the *utilitarian* (that is, a supporter of Utilitarianism) has in mind. In this context, the term means something more like *the ability to produce goodness.* So, an action's utility is a function of its ability to bring about goodness. To paraphrase the words of the famous 18th-century British philosopher Jeremy Bentham (Figure 3.1): *an action is to be approved or disapproved of according to its tendency to increase or diminish happiness.* This is known as the Principle of Utility.

According to this view, whatever you do, you should try to bring about more good than bad. Put another way, in order for your actions to be morally right, they need to maximize *goodness.* Two points ought to be made here. For one, notice that the *consequences* of a person's actions are what determine whether their action was right or wrong. For another, notice that goodness is a fairly broad notion that could be explained in all sorts of different ways. Let us say more about both of these points.

Utilitarianism is a view that focuses on consequences in evaluating rightness. In the final analysis, motives do not really matter. Neither do the particular actions you perform. All that matters to a utilitarian is that your actions result in the greatest overall balance of good over bad. To illustrate that point, consider the vignette below.

VIGNETTE The Backpack Thief

A man walks into a bus station and sits down. On one of the seats near him rests a backpack. Nobody is sitting near it. It appears to have been left behind. The man waits for a while to see if anyone returns to the backpack, but after an hour, nobody has. So, the man stands up, strolls over to the backpack, casually picks it up, and smoothly exits the bus station. Anxious to discover what he has taken, he looks for a place where he can be alone. He spots a large, vacant parking lot a fair distance from the bus station. Once there, the man opens the backpack and discovers… a ticking time bomb! And according to the timer, it is set to explode in 10 seconds! Naturally, the man leaves the backpack and runs for safety. Ten seconds later, the devastating bomb explodes.

FIGURE 3.1 Professor Brei (right) and Professor Bentham (left), the founder of Utilitarianism, in the entrance to the Student Centre at the University College London. (Photo (4 July 2022) courtesy of Andrew Brei.)

Now, how ought we to evaluate this man's actions? If we were to focus on the man's motives, we might blame him for acting greedily and deceptively. If we were to focus on the act itself—that is, the act of taking something that does not belong to you—we might condemn him for stealing. But for the utilitarian, neither of those truly matters in our moral assessment. All that matters are the consequences. And in this example, the consequences were about as good as they could have been. Had the man *not* stolen the backpack, the bomb would have detonated in the bus station. But detonating in an empty parking lot meant minimal damage and no loss of life. Compared to the alternatives, that is a very good outcome. And so, from a utilitarian perspective, the man in this example ought to be praised. His action maximized good and minimized bad. He did the right thing.

Perhaps it strikes you as odd that we would praise somebody who, acting out of greed and selfishness, committed an act of thievery. Rest assured, that strikes many people as odd, including some utilitarians. For those utilitarians, adjustments to the theory need to be made in order that acts of stealing (or worse) would not gain utilitarian approval. But according to the classical version of the theory that we are considering here, the backpack thief ought to be congratulated for having maximized the good.

There is that magic word again: *good*. Because we are doing philosophy here, we ought to be careful to define our terms. Good can mean a number of different things, depending on which view or thinker one consults. But according to the utilitarian theory under consideration, good is equivalent to happiness or, what amounts to the same thing, pleasure. Maximizing goodness means bringing about as much happiness (pleasure) and as little unhappiness (displeasure) as possible in any given situation.

Now that those two points have been clarified, another ought to be made. In order for an action to be right, it needs to maximize happiness to the greatest extent possible in any given situation. (Exactly how much happiness can be maximized will vary from case to case, of course.) But whose happiness matters? According to the utilitarian, we need to consider our own happiness *as well as* the happiness of those affected by our actions. To focus narrowly on our own happiness would be to act in light of a different moral theory: Ethical Egoism [3]. According to that view, an act is right if its consequences involve promoting one's *own* interests, with *no* direct regard for anybody else's. But Utilitarianism requires a more comprehensive appraisal of how actions affect happiness.

Consider, again, the backpack thief. As we indicated above, the utilitarian would praise that person for having acted in a way that brought about as much happiness (and prevented as much unhappiness) as could reasonably have been expected. Of course, it is likely that the thief was unhappy to have stolen nothing of value. But he was not the only person affected by his action—everybody in the bus station benefitted from his action, and a great deal of unhappiness and displeasure were prevented. The important lesson here is that, according to the utilitarian, doing the right thing will not necessarily make *you* happy. Your happiness is just as important as every other individual's happiness—no more, no less.

Naturally, much more could be said about the Utilitarian theory. But for our purposes, it will suffice to remember the following three points:

1. We are talking about a moral theory that focuses on consequences in order to determine right and wrong.
2. The consequences of interest are those affecting happiness and unhappiness.
3. You need to consider both your happiness and that of the people affected by your actions.

Implementation

If a moral theory is worth any attention, it has to be of practical use. It has to help us make the right choices. Utilitarianism can indeed be put to use in this way—and not only to life in general but also to the profession of engineering more specifically. As you will see in later chapters, the Principle of Utility can be used to assess and offer guidance in all sorts of situations that engineers typically face. All one has to do is (1) consider the possible actions one might perform, (2) imagine everybody who would be affected by the actions and how happy or unhappy they would be made, and then (3) opt for the action that would produce the best overall balance of happiness over unhappiness. Simple!

Then again, maybe not. One problem with Utilitarianism is its deceptive simplicity. The theory lends itself to catchy slogans like "the greatest good for the greatest number," or "the needs of the many outweigh the needs of the few," or, indeed, "maximize happiness." Simple though these may be, they do not quite capture the complexity and difficulty involved in envisioning possible courses of action, predicting all of the effects of those actions, and quantifying the imagined reactions to those actions. That is a lot to have in mind! And how is something like happiness supposed to be quantified, anyway? Utilitarianism requires us to determine the amount of happiness caused by our actions—or, alternately, whether our actions bring about high- or low-quality pleasures. But happiness and pleasure are not the sorts of things that can be objectively measured, certainly not to any degree of precision. Such vagueness is concerning.

More worrisome, however, is the utilitarian's exclusive focus on consequences. By restricting their attention to consequences, utilitarians open the door to all sorts of motives and actions that should cause concern. Greed and malice seem like bad ways to be motivated. And bribing, stealing, lying, and cheating seem like bad actions to perform. And yet, if they bring about the right amount (or kinds) of happiness, the utilitarian turns a blind eye to them. That is a problem. Granted, our condemnation of these motives and actions is merely intuitive. That is, we have not proven that these are wrong—we are just supposing that they are. Fair enough… though the burden of proof seems reasonably placed on the person who endorses bribery, rather than the person who condemns it. In any case, ignoring motives and actions should be viewed as a major problem for Utilitarianism.

As you will see in the chapters to come, although moral issues that arise in engineering can be viewed from a Utilitarian perspective, it is far from clear that this is the best perspective to take. Fortunately, there are alternatives.

THE DEONTOLOGICAL PERSPECTIVE

The next moral theory to consider is known as Deontology. (*Deon* is a Greek word meaning "duty," and *-ology* means "the study of.") We will discuss the most famous version of this theory, which comes from the German philosopher Immanuel Kant (1724–1804). According to Kant, there is something deeply problematic with the utilitarian approach to ethics. To illustrate his concern, recall the example discussed above of the backpack thief. Because the consequences of his actions were good (the bomb did not hurt anyone when it exploded), his action would be praised by a utilitarian. But Kant objected. Yes, it is fortunate that things worked out the way they did in that scenario… but the thief should still be blamed. After all, he only "accidentally" saved lives. His real intention was to benefit himself by stealing whatever was in the backpack. And a person should not be praised for being "accidentally" moral.

So, instead of regarding the consequences of actions as the relevant factor, Kant emphasized two other factors: (1) the motivation for an action, and (2) the action itself. In other words, Kant believed that in order to be a good person, one needs to do the right things for the right reasons.

But what are the right reasons? Simply put, we ought to be motivated by a sense of duty. A person should not be motivated by greed, fame, reward, or any other self-interested motive. The only proper motive, believed Kant, is the desire to fulfill one's obligation—to do what's right simply *because it's right*. Kant referred to being motivated in this way as possessing *a good will*.

Naturally, being motivated in that way is fine, as long as we can determine what the right actions are. But how are we supposed to identify the actions we are duty-bound to perform? To help us with that task, Kant articulated a set of principles that can be used to "test" different actions for their moral rightness. Actions pass the test if they are ones we are obliged to perform (and which our motive to do our duty will direct us toward). Actions that fail the test are ones we ought to avoid. We will focus on two of these articulations: the Principle of Universalizability and the Principle of Humanity. But before we explain what these principles say and how they work, a point ought to be made. Although these appear to be two different principles, they are merely two different ways of expressing the same thing. Kant believed that there is one overarching principle that guides us toward right actions—he called that the Categorical Imperative. ("Categorical" means "universal" and an "imperative" is a command.) This imperative can be stated (or formulated) in terms of various different principles, but at their cores, these principles are all saying the same thing. Below we see what they say.

The Principle of Universalizability

Here is one way to state the Principle of Universalizability (hereafter, PU):

> **PU:** Act in ways that adhere to universalizable maxims.

Notice that this is an imperative because it is telling us to do something. It is also categorical because it applies to everybody, not just some. So, this is one way to express Kant's Categorical Imperative. But what does it mean? How are we being told to act?

First, we need to clear up some terminology. A *maxim* is another word for a rule. A maxim also includes the idea of a goal or purpose. What a maxim expresses, then, is a rule regarding a particular action performed for a particular purpose. Thus, the PU is commanding us to follow certain kinds of rules when we attempt to satisfy our goals. But which rules? Only those that are universalizable. In other words, we should only follow rules that would allow us to accomplish our goals *even if everybody else in the world with similar goals also followed that rule*. To illustrate what this means, consider the following fictional vignette.

VIGNETTE Impatient in a Coffee Shop

Suppose I arrive at my local coffee shop one morning and I find a long line of people waiting to order coffee. I really want coffee, and I do not want to wait in line. Then, a thought occurs to me: I could cut to the front of the line and order some coffee. That is an action I *could* perform in that situation. But as a good deontologist, I want to know if I *should* perform it. So, I follow the steps of the PU.

Step 1: I articulate the maxim that the act I am considering (the act of cutting to the front of the line in order to get coffee faster) falls under. That maxim would be something like: *If you want coffee, but there's a line of people waiting for coffee and you don't want to wait, cut to the front of the line.* That is the maxim I would be acting on if I were to perform the action that I am considering.

Step 2: I imagine a world in which everybody acted according to the maxim I have just articulated. In such a world, it is hard to imagine anybody waiting patiently in a line. Instead, it seems likely that a mob of unruly, caffeine-deprived customers would crowd around the counter, elbowing and jostling for position (and coffee). So, if everybody followed the rule that I am thinking about following, chaos would ensue.

Now, perhaps you are thinking: "Wait a minute… deontologists do not focus on consequences!" Quite so. Deontologists do not believe that the consequences of an action determine its rightness or wrongness. "But," you might continue, "at the end of Step 2, we observe that chaos would ensue if our maxim were universalized. And 'ensue' is another word for 'result,' which is another word for 'consequence.'" Again, that is correct. And this would be an embarrassing situation for the deontologist… if not for the fact that there is a third step to this principle.

Step 3: I need to ask myself whether or not my goal of acquiring coffee is achievable in a world where everybody acted according to my imagined maxim. In other words, would cutting to the front of the line make it easier for me to get my coffee? And of course, the answer is: *no!* Fighting my way through a mob is not an efficient, reliable way of acquiring coffee. My goal would actually be frustrated, not supported, by my choice to endorse the act of cutting to the front of the line.

At the end of these steps, we see a contradiction, which could be stated in the following way: *I am going to achieve my goal by acting in a way that would make my goal unachievable.* If I cut to the front of the line, I am putting my stamp of approval on that action. In effect, I am saying that it is the sort of action that everybody should perform. But if that were to happen, then I would not be able to accomplish my goal. And the contradiction doesn't stop there. Not only does cutting to the front of the line make my goal harder to achieve—it also makes it *logically impossible.* For, if everyone were to cut to the front of the line (as I envision myself doing), there ceases to *be* a line. Because the maxim explicitly requires there being a line to cut, the maxim is self-contradictory and self-defeating.

Kant believed that there is something irrational about attempting to accomplish a goal by performing an action that would in fact prevent me from accomplishing my goal. It is that irrationality—*not the consequences*—that indicate the wrongness of cutting in front of the line to get coffee. Thus, the PU helps us see that cutting to the front of the line in order to acquire coffee would be wrong. And it can do the same for any similar action.

But suppose I wanted to resist that result. Could not I simply insist that *only* I should be allowed to cut to the front of the line, while everybody else waits? In such a world, I would surely be able to accomplish my goal of acquiring coffee more quickly than by waiting in line (assuming the barista did not ignore me). I could simply regard myself as the exception to every rule. Fine?

"No!" Kant would reply. None of us is any more or less morally significant than anybody else. My interests are important, it is true—but no more (or less) important than anybody else's. And because we are all equally morally important (for reasons that we will consider in a moment), none of us should imagine ourselves to be the exception to any moral rule. To think so would be arbitrary, unjustifiable, and irrational. This is an important lesson to remember.

The Principle of Humanity

This articulation of the Categorical Imperative (CI), the Principle of Humanity (PH), could be expressed as follows:

PH: Always treat humanity as ends, never merely as means.

The Principle of Humanity relies on a distinction between ends and means. Generally speaking, *ends* are things that we value for their own sake, while *means* are valued because they allow us to achieve ends. The difference here is between things that have *intrinsic value* and *instrumental value*—that is, value based on what something *is* versus value based on what something *is useful for.* The PH instructs us to treat persons (others and ourselves) in ways that respect their intrinsic value, and not in ways that reduce persons to objects or things to be used.

Of course, sometimes we are useful to one another. If you can pardon another coffee-based example, consider my act of purchasing a coffee at a drive-through window. The person who takes my order, prepares my coffee, and hands it to me with a smile through the window is certainly useful to me. But if I take the coffee and gruffly drive away without a "thank you," I would be failing to remember that I was just helped by a person whose value goes far beyond their usefulness. They deserve *respect* by virtue of their humanity. So, even in cases where we are useful to one another, we need to remember that we are all valuable as ends and should be treated accordingly.

As for the basis of our value as ends, Kant points to two features: *rationality* and *autonomy*. We have the cognitive capacity to reason our way through questions and problems. Importantly, we have the rational powers to understand, apply, and act in light of the CI. And we have the freedom to direct our actions according to our will (called "agency"). A deontologist like Kant would regard anything with these features as intrinsically valuable. As for those lacking these features, a more complicated story would need to be told concerning their value. But that story goes beyond the scope of this textbook on engineering ethics. For our purposes, it will suffice to remember that to treat persons as objects, as things to be used, violates the PH because that sort of treatment is at odds with the undeniable value that persons possess.

Implementation

In order to put the deontological theory to work, all an engineer needs to do is think about an action and run it through the tests that the Categorical Imperative provides. Suppose, for example, that I am considering lying to one of my company's clients. The client, let's say, wants to know if the equipment they just purchased from my company was constructed entirely from parts made in the USA. I know that the only non-US parts are two bolts that were made in Belgium. I also know that these bolts are superior in quality to corresponding US bolts and that the Belgian bolts are inaccessible to the regular user of the equipment. The customer will almost certainly never see them. So, I *could* lie... and likely get away with it.

Would such an act pass the PU test? Well, first we need to articulate the maxim in play here. Remember, a maxim expresses both a goal and the action proposed for achieving that goal. In this case, imagine the maxim to be this: *If you want to keep a client happy, tell them what they want to hear... even if you have to lie.*

Next, imagine a world in which every engineer acted in this way. With that much deception going around, it is reasonable to imagine that clients would become quite suspicious and wary of the companies they purchase from. And in such a world, telling the client what they want to hear might not be so reliable a way of making clients happy. So, because my goal would not be achievable (or, at any rate, would become harder to achieve) by way of my proposed action, the PU shows me that I ought not to perform that action.

Again, it is no good thinking that you are the only exception to the rule. If you think that lying to clients is something you should do, while it would be wrong for everybody else, then you are being irrational.

As for the PH, the results are the same. Lying to someone constrains their autonomy, which amounts to an act of disrespect. By lying about the foreign-made bolts (when I know that my client values products made in the USA), I am taking away my client's ability to make an informed decision. I am using my client as a means to my ends (say, of staying on schedule and avoiding conflict). By withholding what I know to be pertinent information from my client, I am failing to show them the respect they do deserve. Thus, the act in question violates the PH.

The same assessment could be made for any action an engineer might imagine, providing insight into what our duties are and which actions we ought to avoid. Of course, Deontology is not a theory without issues. Difficult questions about how to formulate maxims and how to account for the inherent value of non-rational persons, for example, require answers. Knowing, however, that we will not be able to resolve those tricky issues here and that we can proceed well enough with the theory as presented above, we turn instead to another moral theory... and to another approach to doing ethics.

THE VIRTUE THEORY PERSPECTIVE

Near the beginning of this chapter, we noted that there are two basic approaches one might take when doing ethics. One is to focus on actions, and we have just seen two examples of that approach. Now we turn to the other, which involves focusing on character. In making this shift, we concern ourselves less with the question *what should I do?* and more with the question *who should I be?* From this perspective, we aim to understand what it means to be a good person and to live a good life. Clearly, understanding these will shed light on what it means to be a *good engineer...* or rather, an *engineer who is good*. (This distinction became obvious at the end of the section on The Diesel Emissions Scandal in Chapter 1.)

THE NATURE AND SIGNIFICANCE OF VIRTUES

Our examination of Virtue Theory will focus on its most prominent champion, the ancient Greek philosopher Aristotle (384 BCE – 322 BCE). As the name suggests, Virtue Theory is a moral view that emphasizes virtues. (Although Aristotle discussed two types of virtues—moral and intellectual—we will focus only on moral virtues.) Think of virtues as good ways to be. Virtues are character traits that a person can develop and, through practice, come to possess. Aristotle would have identified traits like courage, justice, and temperance as virtues. Other virtues include generosity, honesty, loyalty, humility, and patience. These are ways we can become and, according to the virtue theorist, ways we should become.

In order to better understand virtues, note that they are (in Aristotle's words) "...a mean between two vices, one of excess and one of deficiency" [4]. As we explained in Chapter 1, every virtue exists on a spectrum between two extremes. These extremes are called vices. To illustrate this, think about the character trait *courage*. It is a good trait to possess. It inclines one toward the right sorts of behavior in situations that are scary or challenging. But it is possible to push things too far, to rush into battle when there is no chance of winning. To display bravery in the face of overwhelming odds is not courageous—it is a vice of excess known as *foolhardiness*. Of course, displaying too little courage is a vice as well—a vice of deficiency known as *cowardice*. Neither of these traits is good, and both exist at opposite ends of the spectrum on which courage lies in the middle.

Another way to think about this is by focusing on fear [5]. In dangerous situations, fear is a natural response. To have no fear in the face of danger is no good thing. Neither is having too much fear. And to be oblivious to the fact that there is cause for fear is ignorance—which ought not to be confused with courage. So, recognizing danger, feeling the appropriate amount of fear, and responding with the correct action is what the virtue of courage involves. When one recognizes danger but then displays a disproportionally low (or high) amount of fear by acting rashly (or cowardly), one displays a vice rather than a virtue.

Now, we ought to explain what we mean when we say that a virtue is good (or that a vice is bad). The goodness of a virtue can be viewed from both a practical perspective and a conceptual perspective. Practically speaking, we need other people around us so that we can practice being virtuous. An essential part of being human, according to Aristotle, is being part of a society. We are social and rational animals. As such, we ought to cultivate traits that help us get along with others—and that make it easier for others to get along with us. Being generous, loyal, just, honest, and humble help a person forge and maintain friendships. But a person exhibiting any of the vices relating to these virtues would be a difficult person to get along with. Practically speaking, virtues are good because they allow us to remain in a context of other people, who we *need* in order to practice virtuous behavior.

From a more conceptual perspective, what makes a character trait good or bad is how effectively it allows us to function properly as human beings. Again, our need of others and of friendship is essential to our being human. Said Aristotle, "... [friendship] is a thing most necessary for life, for no one would choose to live without friends, even if he had all other good things." [6]. Having friends engages the social part of our nature, without which we wouldn't *be* human. As such, whatever makes us better social beings automatically makes us better human beings.

That is all somewhat abstract, so allow us to illustrate by way of a simple analogy. Consider an acorn. (Trust us... this is instructive.) If an acorn is to *excel at being an acorn*, then it needs to grow and develop into an oak tree. But in order to do that, it needs a few things. It needs water, sunlight, and the nutrients it can get from soil. We might well say that water, sunlight, and soil are *good* for an acorn—good, in the sense that they are what it requires if it is to excel as the kind of thing that it is. Of course, an acorn needs the right amount of sunlight and water and nutrients. Too much or too little of these goods would not help the acorn to flourish. But, if the acorn gets what it needs, it will become an excellent oak tree.

Similarly, if a human is to *excel at being human*, she needs (among other things) the character traits that will make her function well among others. She needs the moral virtues. Possessing any of these traits in a deficient or an excessive way would not contribute to a person's excellence. That would be analogous to an acorn getting too little (or too much) water. Thus, the virtues are good because they make it possible to live in accordance with human nature—they are a part of what it means to be human.

Practice, Practice, Practice

With a better understanding of what virtues are, we ought to discuss how they can be acquired. And it is worth emphasizing that virtues are traits that need to be developed. Nobody is born generous or loyal—traits like those have to be cultivated over time. But how?

The short answer is: *practice*. In order to become virtuous, a person needs to identify opportunities for acting in virtuous ways. Then, one needs to respond to those opportunities by acting in a specific way (for instance, giving a $10 bill to a needy person to practice generosity; keeping a friend's secret to practice loyalty; limiting oneself to one beer to practice temperance; and so on). And after one has acted in a specific way, one needs to reflect on how effectively her action avoided extremes and reflected a moderate response (perhaps $10 was too little; perhaps keeping the secret harmed others; perhaps one strong beer was enough to make one dizzy; etc.). By repeatedly performing virtuous acts, one can gradually develop the *habits of character* one needs in order to be a virtuous person.

The distinction just mentioned—between virtuous actions and virtuous persons—is important. Naturally, the virtue theorist is concerned with more than merely performing the right actions. She is concerned with the character of the person performing the action. So, we might wonder how long one needs to perform virtuous acts before one becomes virtuous. Aristotle offers as much guidance here as he can:

JUST AND TEMPERATE ACTIONS

"Actions, then, are called just and temperate when they are such as the just or the temperate man would do; but it is not the man who does these that is just and temperate, but the man who also does them as just and temperate men do them. It is well said, then, that it is by doing just acts that the just man is produced, and by doing temperate acts the temperate man; without doing these no one would have even a prospect of becoming good." [7]

In this passage, Aristotle uses justice and temperance as examples of virtues. And his point is that performing a virtuous *act* should not be confused with being a virtuous *person*. It is true that in order to become a virtuous person (which is the goal for a virtue theorist), one needs to perform virtuous actions. But a person is not virtuous until performing those actions has become second nature—a part of their character. (This issue is further addressed as *A Central Ethical Question* in the Epilogue.)

Let us illustrate this point with a musical analogy. Suppose I wanted to become a saxophone player. I would first need to acquire a saxophone. Then, I would need to learn how to attach the reed, how to hold the instrument, and how to operate the various different parts. At first, my actions would be very deliberate and methodical. I would look at a note on a sheet of music, figure out how to arrange my hands and fingers according to the note indicated, and then position my jaw and mouth so as to create the correct situation (*embouchure*, to use a term of art) for blowing into the mouthpiece. Now, suppose I did all that and, on my first attempt, produced a clear, strong, pure note—a C, let's imagine. Am I now a saxophone player?

Certainly not! While it is true that I have done something that saxophone players do, it would be much too hasty to say that I am a saxophone player. Performing the action in question (playing a C) was, for me, very mechanical and calculated, while a true saxophone player would perform the action almost without thinking. After much practice, a saxophone player acquires the habits of musicianship that result in the right notes (and tempo and tone and all the other elements of good saxophone playing) *without* the need for deliberation. They just know when C is the right note to play. Similarly, after much practice, a virtuous person cultivates the habits of character that bring about the right actions at the right times *without* the need for calculation. They just know what to give, when to stay silent, and how much to imbibe. At some point, being virtuous just comes naturally.

So, how long does it take to become virtuous (or musical)? Frustratingly, perhaps, the answer is: it depends. The amount of time it takes to develop habits of character depends on lots of factors, including the influences and examples a person has around them, the opportunities available to a person for practicing virtuous behavior, and the degree to which a person is able (and willing) to reflect on their actions. Everyone is capable of becoming virtuous, just as everyone (in principle) is capable of becoming a saxophone player—although the process can be hard. We will all face our own unique set of challenges and opportunities on the road to virtue. And for that reason, nobody could offer the precise steps or fixed timeline for becoming virtuous. All one could say for sure is that cultivating virtues is an essential part of being an excellent human.

Implementation

As you have already recognized, Virtue Theory does not lend itself to the sort of step-by-step approach to being moral that we found in Utilitarianism or Deontology. That is *not* to suggest that those two action-focused theories are simple—it is merely to suggest that Virtue Theory, with its focus on character, involves a degree of vagueness. It is the sort of moral theory that requires quite a bit from us in terms of implementation. But why should we expect anything else? Nobody but *you* could make you into a saxophone player, and nobody but *you* could make you into a good person (or, indeed, an engineer).

Aristotle gives us the blueprints for being moral, and it is up to each of us to make those plans into reality. In an analogous way, architects and designers create blueprints that engineers and contractors bring to life in ways that are responsive to the unique circumstances of each project. In the moral case, the blueprints Aristotle gives us include some clear, objective truths about what it means to be human and, in light of that, what we need in order to be *excellent* humans.

IEEE CODE OF ETHICS

We now turn from moral theory to moral code. The difference between the two is that while moral theories offer explanations and procedures for determining right and wrong (or good and bad), moral codes simply present a set of rules for one to follow. Moreover, while moral theories relate to our lives in a general social context, moral codes often involve a narrower context—a corporate or engineering context, for instance. The two are related, of course. Any moral code worth our time will be firmly grounded in some moral theory or other. The rules it presents will be the results of a moral principle or framework. Without a solid foundation in moral theory, what would be the point of creating—let alone following—a moral code?

There are countless moral codes (which may also be called *codes of ethics* or *codes of conduct*). They can be simple and straightforward or complicated and detailed. They can be found in companies, educational institutions, governments, and religious traditions. They can also be found in connection with various different professions. Whether one works in education, finance, healthcare, hospitality, journalism, law enforcement, or some other profession, one is almost certain to be held to the standards articulated in a moral code. This is certainly the case for the engineering profession, as well. The National Society of Professional Engineers (NSPE) has a code of ethics [8], as does each specialization within engineering (aerospace, chemical, civil, electrical, industrial, mechanical, nuclear, software, etc.). Here, we will take a closer look at the code of ethics for the Institute of Electrical and Electronics Engineers (IEEE), the full text of which you can find in Appendix B. The early origins of the IEEE Code of Ethics date back to 1906. That year, the president of the American Institute of Electrical Engineers (AIEE) delivered a speech, which eventually led to the adoption of a "Code of Principles of Professional Conduct" in 1912.

As an example, what does the IEEE Code of Ethics (hereafter, *the Code*) have to say about dishonesty, which is a significant ethical concern? According to Section I of the Code, engineers are expected to "uphold the highest standards of integrity, responsible behavior, and ethical conduct in professional activities." And, as I.5 states, those standards entail honesty. Therefore, I.5 is worth quoting here: engineers agree "to seek, accept, and offer honest criticism of technical work, to acknowledge and correct errors, to be honest and realistic in stating claims or estimates based on available data, and to credit properly the contributions of others." This commitment to honesty is indeed fairly evident.

An important note concerning the current IEEE Code of Ethics is the emphasis on the safety, health, and welfare of the public. An engineer's obligation regarding these is *paramount*—of the highest importance—compared to all of the other obligations spelled out in the Code. The unique significance of this obligation has to do with the nature of the engineering profession. Engineers, generally speaking, have the tools and abilities to make great impacts on the world around them. As we know, those impacts often improve lives. But because engineers also have the ability to cause great harms, they must, above all else, avoid activities that would threaten the lives and wellbeing of people. On this, many wise people throughout history have agreed: "With great power comes great responsibility."

Another important note regarding the IEEE Code of Ethics is the significance of the natural environment, which arises (in this context, anyway) because of its importance regarding human health and wellbeing. Humans depend in many ways on a natural environment that is suitably clean, stable, and diverse. And because damages to the environment can result in damages to humans, engineers must remain vigilant regarding their impacts on nature.

As for identifying the ethical underpinnings of the IEEE Code of Ethics, that is to some degree a matter of interpretation. A utilitarian might focus on the negative consequences of damaging nature or harassing coworkers, and, accordingly, condemn those kinds of actions. Then again, those actions (along with every other possible action) would not be wrong *per se*… only when they fail to maximize goodness and minimize badness. The limitations of Utilitarianism have been discussed above, and they will be reemphasized in later chapters. For now, we suggest that the authors of the Code did not have utilitarian calculations of happiness and unhappiness foremost on their minds when writing and revising their code.

A deontologist might focus on the parts of the IEEE Code of Ethics that deal with obligations regarding the treatment of others. The Code instructs engineers to treat persons with respect, which is something that a deontologist like Kant would wholeheartedly agree with. Moreover, the Code hints at the intrinsic value possessed by persons when it *proscribes* (forbids, prohibits) discrimination and harassment. And it seems to include the sorts of imperatives that a deontologist would support. All in all, one could do worse than to regard the ethical foundation of the Code as fairly deontological.

And yet, the language used throughout the IEEE Code of Ethics seems to fit most naturally with the Virtue Theory. Terms like *integrity*, *responsibility*, *honesty*, and *fairness* indicate something about the character of an engineer who is good. And the very notion of committing oneself to a set of standards in support of a certain quality of life indicates the *personal* nature of the Code. It speaks more deeply than to the merely *performative* aspects of being an engineer. It is a set of rules inspired by the importance of living in community with others and of committing oneself to being a certain kind of person.

Think about what *integrity* means. From an engineering perspective, we might say that the integrity of a system, machine, or structure depends (among other things) on the presence, connections, and quality of all the parts necessary for its proper functioning. Similarly, for a person to have integrity, it is necessary that all the "pieces" are in place and sufficiently developed. A traditional internal combustion engine, for example, cannot do what it is meant to do without a functioning spark plug. Thus, we say that a spark plus is an integral part of such an engine. And a person cannot live up to her potential without having cultivated virtues like courage, temperance, and generosity. Virtues, we should say, are integral parts of a human being. So, the IEEE Code of Ethics appears to be firmly grounded on a virtue theoretical account of how to be a good person and, by extension, a *good engineer*—or, indeed, an *engineer who is good.*

SUMMARY

From our discussion of Moral Relativism, we learned that simply describing what communities or cultures accept does not count as doing ethics. In addition, we noted that moral theories are accounts of right and wrong (good and bad) that provide guidance, but which do not themselves determine right or wrong (good or bad). They highlight important factors to consider in making moral judgments, and they offer principles in order to explain and justify those judgments.

From our discussion of Utilitarianism, we learned that one way to do ethics involves focusing on the consequences of our actions. If those consequences involve producing the greatest overall balance of happiness over unhappiness for everyone affected by the action, then (says the utilitarian) that action is right. Of course, this way of thinking permits a number of questionable actions and judgments, and so we considered alternatives.

From our discussion of Deontology, we learned that the motives behind our actions are important, as are our actions themselves. We ought to be motivated by our sense of duty and we ought to only perform actions that we are, in fact, obliged to perform. In order to determine which actions are obligatory, we have two formulations of the Categorical Imperative—the Principle of Universalizability and the Principle of Humanity. Both instruct us to act in certain ways and to avoid the irrationality of regarding ourselves as exceptions to the rules.

From our discussion of Virtue Theory, we learned that character (rather than actions) can be the main focus when we do ethics. Knowing what sorts of character traits we ought to develop requires knowing what a human being is. When we understand that an essential part of being human involves being social, we can see the importance of virtues. Having a virtuous character is important from both a practical perspective and a conceptual perspective. We need others to be comfortable around us so that we can continue to practice virtuous behavior (like loyalty, honesty, and generosity). And we need to live in harmony with others so that we can live an excellent human life—that is, a life that involves living up to our potential as social, rational animals.

Finally, in our discussion of moral codes, we learned that codes of ethics are lists of rules supported by moral theory. In the case of the IEEE Code of Ethics, that support comes most evidently from Virtue Theory. The Code offers guidance for engineers who want to establish habits and practices that will transform them into better engineers and good people.

CONCLUSION

To conclude this chapter and prepare you for the next, we offer a bit of ancient wisdom. Plato, one of the most influential philosophers of ancient Greece, in his dialogue, "The Charmides," writes about a conversation between the philosopher Socrates and Charmides [8]. Charmides was one of the more popular and fun-loving youths in Athens at that time, and he enjoyed partying and overindulging in wine. When he complained about the headaches he had every morning, Socrates claimed to know the cure. Charmides assumed that the cure was something simple—nowadays, we might think of a pill (such as ibuprofen) or a supplement. But according to Socrates, it is more complicated than that. A cure needs to reach deep down, to the *source* of the problem. He advised Charmides:

BEGIN BY CURING YOUR SOUL

"For all good and evil, whether in the body or in human nature, originates… in the soul, and overflows from thence, as if from the head into the eyes. And therefore, if the head and body are to be well, you must begin by curing the soul; that is the first thing. And the cure, my dear youth, has to be affected by the use of certain charms, and these charms are fair words; and by them temperance is implanted in the soul, and where temperance is, there health is speedily imparted, not only to the head, but to the whole body." [9]

Well, perhaps that warrants further interpretation. What Socrates meant is that by focusing only on the symptom—Charmides' hangover, in this case—you do not really address the underlying issue. Consequently, the symptom will keep returning. Only by identifying and correcting the ultimate cause of a problem can you achieve permanent relief.

In Charmides' case, the headaches are caused by drinking too much wine. But what causes that? According to Socrates, it is the fact that Charmides lacks the virtue of temperance. This character trait allows a person to behave moderately, with self-restraint and self-discipline. Without it, one is liable to act in extreme ways, either overindulging or underindulging in the pleasures of life. So, by engaging in temperate acts—the "fair words" Socrates mentions—one can develop a temperate character. And once that habit of character is in place, one will be far less likely to overindulge in drink and, hence, suffer from headaches.

However, by pointing to what Socrates said about temperance being the "cure" for hangovers, we are *not* suggesting that every instance of alcoholism is caused by a lack of virtue (i.e., viciousness). Every analogy has its limits, and that surely includes the analogy between ethical and alcoholic hangovers, too.

What is the lesson here? Being moral is not a simple matter, nor is it something that you can be "made into" by some outside force. Being moral involves awareness, insights, and commitments for which each of us is responsible, individually. Addressing unethical behavior in any way other than addressing its *ultimate cause* will fall short of solving the root problem. That includes Moral Relativism [1], Behavioral Ethics (to be discussed in Chapter 4), and any other descriptive approach. Complementary and engaging as these may be, they fail to account for the fundamental truths that ground morality—truths that, to some degree of accuracy or other, are revealed in the study and application of moral theory.

Even so, there is value in knowing more about *how* humans think and behave when the temptation to do wrong presents itself. And there are some practical benefits to be gained from understanding the process by which we give ourselves moral licenses. A commonsense model of this process (remember Figure 2.1) can also be used to explain unethical behavior among practicing engineers. We turn to that interesting topic in the next chapter.

CLASS EXERCISES

1. Create mind maps of the following sections:
   a. The Utilitarian Perspective
   b. The Deontological Perspective
   c. The Virtue Theory Perspective

   Class discussion: Compare the mind maps of different students. What is the educational value of mind-mapping in this case?
2. Evaluate an act of *lying to customer about a product* from the perspectives of Utilitarianism, Deontology, and Virtue Theory. Break up the class into three groups, each assigned one of the three moral theories. Each group will analyze the rightness or wrongness of lying according to their assigned perspective.

Class discussion: Note how the different theories offer different assessments. Is it possible to say that one of these moral theories is *better* than the others in this context? What criteria and arguments would you use to determine such superiority? Is your argumentation, in fact, relying on some moral theory? (In other words, is that question a kind of chicken-and-egg issue, perhaps a philosophical paradox?) Or are you relying on objective facts to support your claims?

3. Ask one of the popular AI chatbots (ChatGPT, Copilot, or Gemini) to offer a moral evaluation of an act of *bribery*. Then analyze and assess the chatbot's assessment.

   Class discussion: Identify the weaknesses of AI-based moral evaluation. What sorts of vagueness or inaccuracies can you find? Do you see AI-based and human-based moral evaluation as competitors or could they be complementary, in the long run?
4. In addition to the IEEE Code of Ethics, what other moral codes are you aware of? Find one and determine which moral theory serves as its most obvious foundation. Why is not it enough to have just a single code of ethics for *all* engineers?

   Class discussion: Compare your findings with other students. Is it really *beneficial* to have different codes of ethics for different fields of engineering?
5. First, study relevant sections of the online document on Moral Relativism [1]. Assume it is the *only* moral theory practiced in your community. Next, as a "trailer," examine the section of Workplace Harassment in Chapter 9, and consider the vignette "You're Doin' Our Job."

   Class discussion: Based on the guidance of Moral Relativism practiced in your community, did Walter do *wrong* when he verbally harassed Kurt in that vignette? Or was Kurt just too sensitive? After all, Kurt was a foreign newcomer to Alabama—deep south and the Heart of Dixie.
6. In classical multi-cultural problems, bridging multiple cultures can benefit everyone involved. Perhaps, this bridging might apply to different moral theories, too. If that were the case, an individual's moral standards could be a *hybrid* of Utilitarian, Deontological, Virtue Theory, and possibly other perspectives. Think about your personal ethical standards. Do you recognize such hybridization in them in varying conditions and situations? (This resembles ensemble methods in computational intelligence.)

   Class discussion: Compare the "hybrid moral standards" of different students. Are there any similar patterns in how these hybrids are formed, and in how dynamic situations are handled with such multi-theory-backed ethical standards? Note: this discussion needs *strong moderation* by the instructor.

REFERENCES

1. C. Gowans (2021), "Moral relativism," in *The Stanford Encyclopedia of Philosophy*, E. N. Zalta, Ed. Accessed: Oct. 18, 2024. [Online]. Available: https://plato.stanford.edu/archives/spr2021/entries/moral-relativism/
2. J. Driver (2022), "The history of utilitarianism," in *The Stanford Encyclopedia of Philosophy*, E. N. Zalta and U. Nodelman, Eds. Accessed: Oct. 18, 2024. [Online]. Available: https://plato.stanford.edu/archives/win2022/entries/utilitarianism-history/

3. R. Shaver (2023), "Egoism," in *The Stanford Encyclopedia of Philosophy*, E. N. Zalta and U. Nodelman, Eds. Accessed: Oct. 18, 2024. [Online]. Available: https://plato.stanford.edu/archives/spr2023/entries/egoism/
4. Aristotle, *Nicomachean Ethics*. H. Rackham, Ed., Book II, Chapter 7, Section 15. Tufts University. Accessed: Oct. 18, 2024. [Online]. Available: https://www.perseus.tufts.edu/hopper/text?doc=Perseus%3Atext%3A1999.01.0054%3Abook%3D2%3Achapter%3D7%3Asection%3D15
5. Acknowledgment: The authors would like to extend their gratitude to Dr. Robert Boyd Skipper, one of the field expert reviewers, for suggesting this shift in focus from *courage* to *fear*.
6. Aristotle, *Politics*. B. Jowett, trans., Book VIII, Chapter 1. McMaster University. Accessed: Oct. 18, 2024. [Online]. Available: https://historyofeconomicthought.mcmaster.ca/aristotle/Politics.pdf
7. Aristotle, *Nicomachean Ethics*. H. Rackham, Ed., Book II, Chapter 4, Section 4. Tufts University. Accessed: Oct. 18, 2024. [Online]. Available: https://www.perseus.tufts.edu/hopper/text?doc=Perseus%3Atext%3A1999.01.0054%3Abook%3D2%3Achapter%3D4%3Asection%3D4
8. "NSPE Code of Ethics for Engineers." NSPE. Accessed: Oct. 21, 2024. [Online]. Available: https://www.nspe.org/sites/default/files/resources/pdfs/Ethics/CodeofEthics/NSPECodeofEthicsforEngineers.pdf
9. Plato, *The Dialogues of Plato: Charmides*. B. Jowett, trans. The Project Gutenberg. Accessed Oct. 18, 2024. [Online]. Available: https://www.gutenberg.org/files/1580/1580-h/1580-h.htm#link2H_4_0003

4 Tools from Behavioral Ethics

Learning Objectives

After studying Chapter 4, you will be able to

- Understand the descriptive, behavior-based approach to ethics
- Use moral and psychological self-licensing to explain moral and other micro-level actions
- Apply moral currencies for gaining high-level understanding of unethical processes
- Describe the dimensions of the corporate ethical virtues model that are negatively related to perceived unethical behavior in real organizations

Behavioral ethics is a study of why people make ethical or unethical decisions, and it belongs to the field of social scientific research. It differs considerably from moral philosophy, which is based on moral theories like those introduced in Chapter 3. Rather than focusing on how people *should* behave in order to be good or do the right thing, behavioral ethics examines *what* people do and *why* they do it [1]. Hence, it provides practical means for explaining—but not evaluating or justifying—unethical actions in various contexts, including engineering. It is important to understand that from the holistic perspective of engineering ethics, moral philosophy and behavioral ethics are complementary and not in competition with each other. They are doing different things—one is normative, the other descriptive. And while behavioral ethics is no substitute for normative ethics (which relies on moral theory), engaging in both is vital to the improvement of community, culture, and self.

In the following, we provide a discussion of behavioral ethics—what it is and what it can offer the readership of this book. We then introduce related tools, including psychological and moral self-licensing [2, 3], to explain the license-granting stage of our commonsense model of (unethical) actions. This text uses these tools to (1) help readers gain insight into the cases presented in Chapters 5–7, and to (2) apply psychological and moral self-licensing as convenient tools to understand and explain everyday decision-making. And it becomes obvious that unethical actions *do not just happen*, but they must first be authorized by the actor himself. This kind of viewpoint establishes a distinctive and novel approach to studying engineering ethics. Finally, a multidimensional model of ethical culture is presented to explain unethical behavior within organizations [4].

DOI: 10.1201/9781003485520-4

BEHAVIORAL ETHICS

Bazerman and Gino define *behavioral ethics* in their review article as: "The study of systematic and predictable ways in which individuals make ethical decisions and judge the ethical decisions of others that are at odds with intuition and the benefits of the broader society" [1]. They see behavioral ethics as a response to the fact that traditional ethics draws little *empirical* attention to how people behave or how their behavior could be improved. Next, we present a sample of research on behavioral ethics, which is particularly relevant to engineering professionals.

The deep roots of traditional "Western" ethics (a.k.a. moral philosophy) reach back to the works of ancient Greek philosophers, such as Plato and Aristotle. But the dawn of behavioral ethics era is much more recent; it probably began when David Messick and Ann Tenbrunsen's (eds.) book *Codes of Conduct: Behavioral Research into Business Ethics* was published in 1996 [5]. In those days, three types of theories that individuals use in making moral decisions were identified [1]:

- Theories about the *world*
- Theories about *other people*
- Theories about *ourselves*

This trinity of theories has shown to be useful in understanding ethical decision-making. Furthermore, it was suggested that moral behavior (though *not* moral truth) is malleable and dynamic: even people who strongly appreciate morality may not act consistently in different situations, and they can exceed their ethical boundaries under case-by-case pressures (such as peer pressure, time pressure, or pressure from management) [1]. For this reason, virtually, everyone may get an ethical hangover every now and then.

Given the role of individuals' psychology in explaining their unethical actions, empirical research has been conducted to explain the *situational* and *social* forces that influence how people make ethical decisions and solve ethical dilemmas—or even polylemmas. This research has discovered that the more opportunity afforded by a situation for people to rationalize their behavior, the more likely they are to behave unethically [6]—thus "opportunity makes a thief." And the environment in which people operate will activate explicit or implicit standards, which in turn affect the tendency to cross ethical boundaries. There may be differences between one's personal ethical standards and those promoted at workplace, for instance. In addition, related work on social rather than situational factors found that the actions of other people can influence one's own actions in the ethics domain [1]. This is especially true when subordinates deal with their ethically or unethically behaving superiors. Thus, "leading by example" can have both positive and negative consequences—depending on the example.

Hence, behavioral ethics provides an approach for students and young professionals to better understand their own behavior, and compare it with how they should behave or how they would want to behave. Bazerman and Gino conclude their review: "We believe that only by reflecting on their ethical failures and the inconsistencies between their desire to be moral and their actual behavior they can rise to the actions (and ethical standards) that their more reflective selves would recommend" [1]. After this general introduction, we are ready to present specific tools that we use later to explain our quasi-factual case examples of the behavior of engineers within their organizations. These tools can also be applied as convenient ethical problem-solving methods.

PSYCHOLOGICAL LICENSING

The desire to do something is a necessary, but not alone sufficient, condition to motivate action. In addition to wanting to do something, one must also feel entitled to do it [2]. Thus, the *Desire* that one experiences initiates a sequential process, which first aims to license the action and then performs it. But if the requested license is *Not granted* due to existing criteria (an individual's other commitments, for example), the action is not allowed. This process is illustrated in the commonsense model of Figure 4.1, where the dashed feedback line shows how the *Action details* can be used for adapting the license-granting criteria for similar desires in the future.

The license here is called *psychological license*, and it covers a wide variety of actions that an individual may encounter. However, if the particular action involves moral behavior, it needs a *moral license* (Figure 2.1), which is a special case of psychological license. We assume that all such licenses are granted by the actor himself, and thus, all the actions to be performed are *self-licensed*. A psychological license can be based on innumerable motivational factors, including the examples listed in Table 4.1, some of which are virtues. And if there are enough motivational factors with sufficient strength, a psychological self-license can be granted to a particular desire—like going hiking.

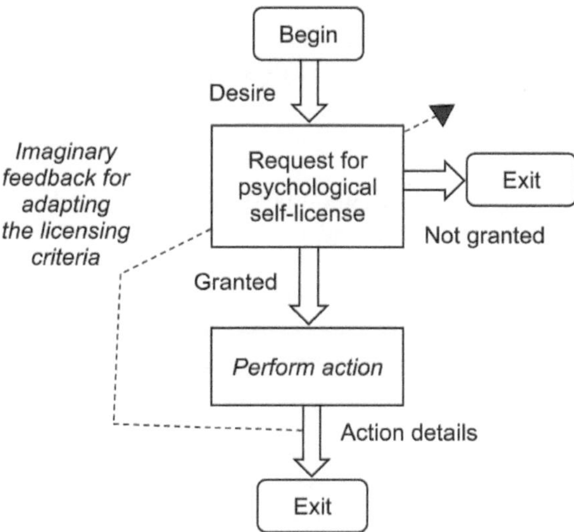

FIGURE 4.1 Commonsense model for licensing and performing actions (note: *excluding* moral actions).

TABLE 4.1
Typical Motivational Factors behind Psychological Licenses

Motivational Factor
Beauty (v)
Curiosity (v)
Economy
Entertainment
Health
Helpfulness (v)
Love (v)
Responsibility (v)
Safety
Wellbeing

Note: (v) represents a virtue.

Miller and Effron define the term *psychological license* as: "The notion that a person has permission to take action or express thoughts without fear of discrediting himself/herself" [2]. Moreover, they point out that when people inhibit their attitude expression because they feel it is "not their place" to speak up, they lack what is known as *psychological standing*. Such a standing—a type of license—can be achieved, for instance, by belonging to a particular social category, having certain experience that connects the person to the matter in question, or having some involvement in the matter. Our actions must ultimately be authorized by us—and no one else.

Below, we present examples of psychological self-licensing to emphasize the fact that things do not "just happen," but, strictly speaking, are self-licensed by the actor.

Things Do Not Just Happen

Let us first go through two cases and the associated psychological licenses. The fictional vignette below is intended to demonstrate the core message of this section.

VIGNETTE Self-Licensed or Not?

FIRST CASE

Gunther was downhill skiing on a sunny mountain slope in late spring, making his way down a natural slope outside a ski resort. His speed picked up and he felt free. But suddenly, after a thrilling jump, there was a largish spot without any snow—the hot spring sun had melted the snow away. And Gunther could not pass the gravel spot or jump over it. Thus, he crashed at high speed.

SECOND CASE

Gunther was riding his mountain bike in an Alpine valley. The trail was rather easy and he enjoyed the beautiful scenery. Suddenly, he heard rumbling and the ground shook violently for a few seconds—an earthquake! And shortly, he heard vigorous noise from the top of the mountain; a huge block of glacial ice was sliding downward. Gunther tried to get out of its way but was not fast enough.

Were these two unfortunate actions licensed by Gunther? An immediate answer might be: no, both were accidents. But let us see. In the first case, Gunther had granted himself a psychological license to go downhill skiing in that area. The main motivational factors for his license were entertainment and wellbeing, but safety was important, too. However, it is not uncommon for sunny mountain slopes to have snow-free spots in late spring. Gunther must have known that, but he either ignored such a scenario or considered its probability of occurrence to be negligible—maybe he was a bit careless. Therefore, the crash that occurred was actually predictable, albeit with a relatively low probability. Strictly speaking, it was a self-licensed outcome. Clearly, there was a problem with *uncertainty management* of safety when Gunther created his psychological self-license. In fact, the same applies to most so-called accidents.

In the second case, Gunther had given himself a psychological license to go mountain biking in a geologically stable region, which had not experienced earthquakes for at least 100 years. Nor were there any early signs of seismic activity. The mild earthquake and its consequences were unpredictable. Thus, what happened to Gunther was not a licensed action—it just happened—it was a pure accident. And the motivational factors behind Gunther's psychological self-license had been health and wellbeing, as well as safety.

The fundamental difference between these two cases is that the first case was *predictable* and the second was *unpredictable*. In the first case, the psychological license itself had problems of uncertainty management in terms of safety. But the second, sporadic case does not fall under the category of inadequate uncertainty management. Such an *extended licensing* principle can be formulated in the traditional terms of logic as shown below.

Extended licensing: If action Q is licensed by actor x, then outcome W, which is either a deterministic or stochastic consequence of Q, is also licensed by x; this excludes sporadic consequences. Here, *sporadic* means an event that occurs randomly at very irregular intervals and sparsely, while *stochastic* means an event having a random probability distribution.

If we next omit the stochastic consequence option, we get the more familiar *basic licensing* principle. And according to this relieved principle, both of the above cases would have been pure accidents.

Basic licensing: If action Q is licensed by actor x, then outcome W, which is a deterministic consequence of Q, is also licensed by x.

Basic licensing corresponds to the ordinary logical implication, and extended licensing can be seen as its stochastic counterpart.

We emphasize that only unexpected natural events or acute disease attacks "just happen," and their immediate consequences leading to whatever outcomes are therefore not licensed. Practically, everything else a person encounters is thus self-licensed (either implicitly or explicitly), and the individual is to some degree responsible for it. Of course, such extreme conditions as psychosis or Alzheimer's disease render an individual incapable of granting rational self-licenses.

Although our extended licensing principle may appear odd and overly strict, it forms a sound basis for the critical and precise study of moral decision-making later in this chapter. How come? Because it is important to take into account the *unavoidable* aspect of stochastic uncertainty, whenever assessing the consequences of contemplated unethical actions for others and oneself.

Connection to Personality

The threshold for granting self-license is sometimes related to an individual's willingness to take risks. And the eventual outcome of the action performed—successful or unsuccessful—could be used to adapt his risk-taking threshold—as we know, "to err is human."

More than a hundred years ago, Swiss psychiatrist Carl Jung introduced two different personality types, *extrovert* and *introvert.* According to Jung's contemporary, Maurice Nicoll, an extrovert "flows out into action easily," and "comes in contact with life eagerly, spontaneously, without preparation or plan" [7] (p. 139). On the other hand, an introvert "is thoroughly aware of his inner life and is a keen and serious critic of himself" [7] (p. 148). Intuitively speaking, it seems obvious that pure extroverts are more prone to taking risks than pure introverts. However, most real personalities lie somewhere between these extremes. Personality is, indeed, an important driver for the psychological basis of self-licensing. And, for example, the ability to empathize also belongs to the same category—it affects one's criteria to grant a psychological self-license.

To conclude, a licensed person is a disinhibited person; one for whom a certain psychological barrier has been removed [2]. Explicit *awareness* of the existence of psychological license helps to develop an individual's ability to understand and regulate his actions. And the ultimate result can be a prudent "pause-and-think" approach to making critical choices. After understanding the principles of psychological licensing, we are now ready to explore its important special case, moral licensing.

MORAL SELF-LICENSING

Moral Credentials

Sometimes people have a desire to act in ways that may call into question their moral values. In such tempting cases, they must first present themselves with adequate reasons for unethical behavior—this was discussed already in Chapter 2. These reasons collectively constitute a moral self-license (Figure 2.1). Effron demonstrated that people in different situations are skillful at convincing themselves that they have a moral license to give into temptation [3]. They can construct their license by

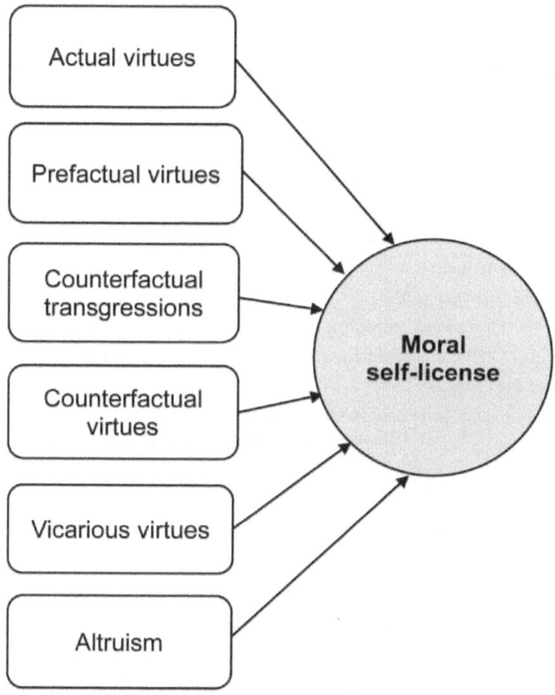

FIGURE 4.2 Composing a moral self-license from possible sources of moral credentials [3].

applying different strategies and using a variety of moral credentials, including those shown in Figure 4.2. Fortunately, the notable desire to feel and appear virtuous often prevents people from giving into unethical temptations.

Further, Effron clarifies five of these credential types briefly as follows:

I. Actual virtues: Good deeds you have performed
II. Prefactual virtues: Good deeds you plan to perform
III. Counterfactual transgressions: Bad deeds you declined to perform
IV. Counterfactual virtues: Good deeds you believe you would have performed
V. Vicarious virtues: Good deeds others have performed

We include a sixth (VI) moral credential in Figure 4.2—"altruism," or the desire to put the interests of others before one's own interests. This is a common type of credential in the context of workplace. Table 4.2 lists a sample of possible moral licenses that could, in principle, be used by practicing engineers.

People can also use *advice* as a license element to make ethically questionable decisions, as shown below where an imaginary engineer "was advised to do so by a respected lawyer." And forcing—either implicitly or explicitly—to act unethically is clearly a form of extreme advice. On the other hand, certain unethical behavior can even appeal to *fairness* in a tit-for-tat way [8] (p. 82), such as one's "right to retaliate,"

TABLE 4.2
A Fictitious Set of Elements that Could Be Used in Engineers' Moral Self-Licenses

License Element
He had always done his work very professionally (I)
He saw his company as being under an attack by a customer (VI)
He was advised to do so by a respected lawyer (advice)
He reinforced his earlier lie to keep it alive (reinforcement)
He was practically forced to act so (advice)
He would work really hard next week to completely make up for the wasted work day (II)
None of his team members had ever before cheated their company (V, group)
Solely for the benefit of the company (VI)
They did me wrong and hence I have the right to retaliate (fairness)
This had never happened before, although there had been several opportunities (III)

Note: A roman numeral in parentheses indicates the type of moral credential identified.

mentioned above. Furthermore, Sissela Bok states in her book: "[Often], the first lie 'must be thatched with another or it will rain through'" [8] (p. 25). Such a common *reinforcement* viewpoint is also present in Table 4.2. Participation in a *group* can be a significant motive for lowering one's own ethical standards. It can be very tempting to go along with what a group is doing, even if that involves actions that violate one's own standards. Nevertheless, conformity to group's norms allows members to accumulate idiosyncrasy credits that can later be used to "purchase" rights to deviate, if needed [2]. Thus, a member's faithful history of norm conformity negates the discredit that occasional norm violations would signal to the group. This is similar to trading moral currencies, to be discussed later in this chapter.

In general, resisting morally questionable temptations requires mental effort (self-discipline), which may be a finite resource. And when this resource is exhausted, people are more likely to give into unethical impulses [9]. However, self-discipline is actually a virtue that can be acquired and strengthened through practice.

Assessing Consequences

The model of moral credentials discussed above resembles the current practice of purchasing "carbon-offsets" to compensate for one's emissions of carbon dioxide or other greenhouse gases [2]. So, if purchasing carbon-offsets is sometimes considered as "greenwashing," would moral self-licensing then be a sort of "ethics-washing?" We will return to this relevant question after the following paragraph.

When an actor deceives someone, or a group of actors deceives, for example, the customers of their company, the consequences of the deception are usually only viewed from the perspective of the deceived. That is, we tend to focus on the impacts of the lie on those to whom it was told. But what about the actor or actors himself/themselves? The actor constructed a moral license to give himself permission to behave unethically—he thought "even though such an action might be bad or wrong, it's still acceptable for me to do it." This was a conscious decision (though licenses for minor actions could also be granted subconsciously). But the actor's conscience uses a different procedure to assess the outcome of his self-licensed *unethical* behavior. If the *Action details* in the model of Figure 2.1 are severe enough, the final classification would be "intolerable." Hence, the actor's conscious and subconscious ethical norms are in conflict, leading to an ethical hangover. However, according to our commonsense model of unethical behavior, there is an individual-specific region within *Action details* that the actor's conscience can classify as "tolerable," and therefore, conscience does not always penalize the "mildly unethical" actor.

Let us now return to the question, *is* moral self-licensing a form of "ethics-washing?" Well, we think it is. The more actions one can label "tolerable," the more unethical actions one can commit without any negative internal consequences. The wrongness persists, but "it doesn't hurt." As an extreme case, it has been suggested that psychopaths lack a conscience [10]; if that were the case, everything could be considered tolerable for them.

Additionally, the unethical actor can naturally experience external consequences for his ethically questionable behavior. These consequences are stochastic in nature; they may or may not appear with varying probabilities. And in accordance with the *extended licensing* principle proposed earlier in this chapter, they are also licensed by the actor himself. If it turns out that he lied, for example, his credibility and respect for his word are damaged. Sissela Bok writes wisely in her book: "Even if the liar cares little about the risks to others from his deception, … all [the] risks to himself argue in favor of at least weighing any decision to lie quite seriously" [8] (p. 26). This wisdom holds considerable power by comparing how plausible a justification for a lie seems to the liar versus to the victim. Unfortunately, liars rarely consider such wisdom, and the resulting deterioration of trust within the workplace can cripple even their benevolent aims [11]. And it surely is difficult and time-consuming to restore the lost trust again.

Next, we discuss how challenging it is to take into account the meaningful consequences of planned unethical action—for both others and the actor. When a prospective actor tries to list and comprehend, for example, the potential consequences of his contemplated lie, it is easy to ignore even quite obvious consequences and their non-zero probabilities of occurrence. Often, there are many potential consequences, and their probabilities are hard to estimate. And when these uncertain consequences are put together with the partly uncertain benefits of the lie to the actor, the decision-making task becomes even more cumbersome. In fact, we can see parallels between such stochastic decision-making and belief-based reasoning [12]. Therefore, the majority of actual decisions leading to lying or some other unethical behavior are necessarily subjective or at best semi-objective (a mixture of objective and subjective)—without much uncertainty management. This can, of course, lead to unexpected and unpleasant consequences for the actor himself, even though the moral self-license appeared so strong. Table 4.3 shows a representative subspace of issues related to consequences for unethical actions.

How often have we taken into account these issues *objectively* when contemplating an unethical action? Perhaps never to this extent. And this is only a subspace

TABLE 4.3
A Subspace of the Issues Related to Consequences for Unethical Actions

Issue to Consider
Who are the immediate/secondary victims?
What are the possible short-term/long-term consequences?
How long is the estimated duration of the impact of individual consequences?
Which of the consequences are local and which have a broader impact?
Which of the consequences are deterministic/stochastic?
What are the estimated probabilities of the stochastic consequences?
Do the consequences affect both professional and personal life?

of the entire space of relevant issues. Hence, building on the basis of moral self-licensing is more or less fooling oneself—but we do it anyway. Nonetheless, moral self-licensing is a convenient tool for *explaining* and *understanding* unethical behavior, as well as an important new topic for the teaching of engineering ethics.

MORAL CURRENCIES

To supplement moral self-licensing with another explanation tool, Weisel and Shalvi proposed the system of *moral currencies* [13]. They suggested that in order to justify morally questionable choices—when people desire or are practically required to behave unethically—people engage in a moral calculus in which they regard ethical values and behavior as moral currencies. These currencies can be traded with each other using certain exchange rates. Such ethical and unethical actions can also be temporally distant and independent from each other. And the same action can even have both positive and negative ethical values. An example of such a hybrid-value action is lying, which is certainly a dishonest action, but it can sometimes be seen as a loyal action, too [14].

Considering morality in terms of exchangeable currencies can help understand people's choices when faced with options that promote one value at the *expense* of another (as occurred in the section of The Diesel Emissions Scandal in Chapter 1, where the VW software developers turned the deceptive scheme into reality) [13]. According to this view, ethical values—such as loyalty and honesty—can be seen as moral currencies that can be traded with each other. Hence, the system of moral currencies is a sort of accountant's perspective to ethics.

Balance, Loan, and Payback

Let us imagine an Ethics Bank, where an actor has a personal savings account. He deposits moral currency to his account when his behavior has been virtuous and withdraws currency when he has behaved unethically. But if he is planning an unethical action that would require more currency than his balance would allow, he can borrow a sufficient amount from his bank. Alternatively, he could borrow the necessary amount from one of his friends, for instance—and then he owes his friend. And, of course, all loans must be paid back sooner or later.

In addition, just as ordinary banks pay interest on different accounts, the Ethics Bank could also pay the actor an imaginary interest if his account balance has been positive for a relatively long time. This means that the recent behavior of the actor has been mainly virtuous. Unfortunately, the imaginary interest could promote unethical behavior, because "being good frees us to be bad," as we learned from moral self-licensing [3]. Nonetheless, it is still possible to avoid the temptation—as we all know. Such an imaginary interest-rate principle could also be applied to loans; the longer the payback period, the more interest has to be paid to the bank. Furthermore, if the payback is delayed over and over again, the ultimate result may be an *ethical hangover* ("insolvent" in banking terms). These interest extensions to the moral currencies system are based on our intuitive thinking.

An instructive example of such a bank account holder could be the well-known Robin Hood. According to an old legend, Robin Hood and his band of outlaws stole from the rich people in order to give to the poor. Stealing is definitely unethical and illegal, but giving (that is, being generous) to poor people is considered ethical and respectable. Here, the unethical action is carried out with the *intention* to follow it with an ethical action. Assuming Robin's account at the Ethics Bank was empty to begin with, he first had to loan moral currency to be allowed to steal from the rich. And later, he paid back his loan by giving the stolen goods or money to the poor. Here, the assumed currency values of dishonesty ($-C$) and generosity ($+C$) can be accumulated on each other. As $-C+C=0$, Robin's moral self-image stays clean! This was an altruistic chain of action, but we can interpret the stealing-giving pair of actions as a single meta-action. In general, when the dishonesty is not only altruistic, but also self-serving, the temptation to deceive is especially seductive [13].

Today, Robin Hood's statue is near Nottingham castle, in England. The robber has become a hero—but is this the right message to give the public? Well, maybe it does not matter, because Robin Hood is just an old legend (dating back to at least the 14th century). However, it is *not* wise to praise unethical action in a workplace, although the particular action could be seen as beneficial from the company's point of view. If a liar, for example, is treated as a kind of hero, it gives a contradictory signal to members of the organization. Instead, a recommended approach would be to sit down and discuss seriously what could be learned from this incident. Chapters 5 and 7 analyze two unethical cases in real engineering organizations involving such praise (Box 5.2 and Box 7.1).

Obviously, the levels of honesty and cooperation vary across societies and all types of cultures, and thus, it is unclear whether local conditions and traditions affect the internal trading value of a particular norm, such as honesty [13]. And while the moral currency system provides a quick and straightforward way for gaining a *general* understanding of unethical processes, it shares somewhat similar uncertainty challenges that we discussed above with moral self-licensing. The central question is, how to determine the trading value of individual moral currencies; for instance, how would tactlessness compare with responsibility? Answers to this and other currency-valuation questions are subjective, in the best cases semi-objective. Moreover, the problem of currency exchange rates becomes even more difficult if the currency trading takes place between individuals or over lengthy time periods. Weisel and Shalvi close their article with an open question—they ask whether a person's internal moral-currencies system uses a fixed exchange rate, or an exchange rate that depends on the particular situation [13]? Our intuition tells us that the exchange rate may depend on the situation. For instance, a lie is a lie, but lying to a close co-worker may carry a different exchange rate than lying to a client, reflecting the situational nature of moral-currencies.

UNETHICAL BEHAVIOR WITHIN ORGANIZATIONS

Unethical behavior in various organizations is a widespread phenomenon: the 2013 study by KPMG Forensic of more than 3,500 industry managers and employees in the United States shows that nearly three-fourths of them observed unethical behavior in their organization in the previous year. And more than half of the respondents reported that what they had observed could cause a *significant loss of public trust if discovered* [15]. There is therefore a lot of potential and need for improving the ethical culture of organizations. A company's ethical culture—or the lack of it—affects both the micro and the macro levels, as we will see in the upcoming case-study chapters.

According to that recurring study by the consulting company, the most commonly cited drivers for unethical misconduct in 13 industry sectors were [15]:

- Pressure to do "whatever it takes" to meet business goals (cited by 64% of respondents)
- The organization's code of conduct not being taken seriously
- The fear of job loss over unmet goals
- Being rewarded for results and not the methods to achieve them

The first of these drivers is also involved—either indirectly or directly—in most of the case studies that we present in the following chapters. Moreover, we believe it was behind the Volkswagen diesel emissions scandal that was introduced in Chapter 1, too.

Corporate Ethical Virtues Model

Engineering ethics has clearly two sides: the *individual* side and the *organizational* side. So far, we have mainly considered the individual side; but below, we examine the Corporate Ethical Virtues (CEV) model, introduced by Kaptein, which reveals a little bit of the other side [4]. And a workplace can only be ethically sound if both the employees and their organization are committed to ethical conduct. At the time of its introduction, the CEV model was the only multidimensional model of ethical culture in organizations. Figure 4.3 shows the eight dimensions of that pragmatic model, which together make the phrase "ethical culture" more concrete.

For his empirical research to test the predictive validity of the CEV model, Kaptein first developed eight hypotheses based on the individual dimensions of ethical culture. All these hypotheses were of the form: *the dimension X is negatively related to*

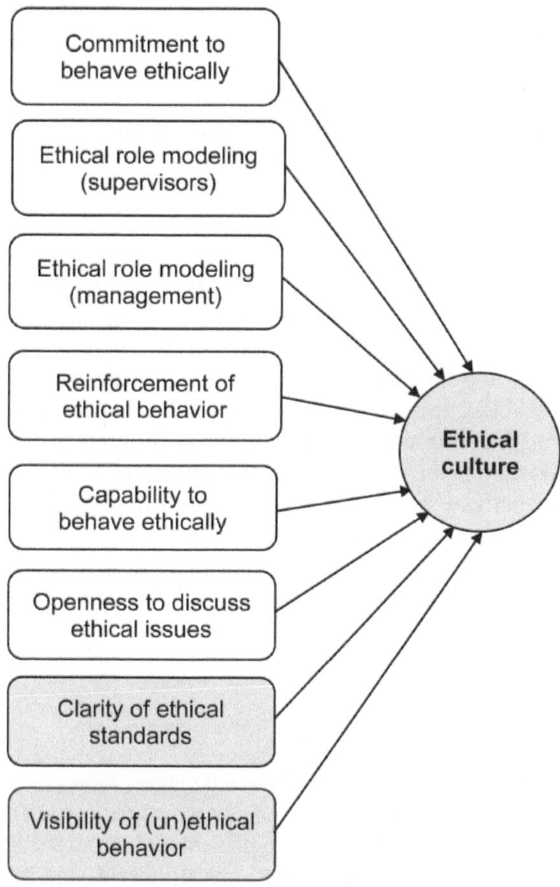

FIGURE 4.3 The Corporate Ethical Virtues model [4].

the frequency of unethical behavior in the workplace. And the comprehensive data were collected from multiple work settings with more than 300 heterogeneous work groups, in organizations of various sizes. A questionnaire with 58 questions was used to measure the dimensions of ethical culture. Almost 16% of the workgroups analyzed reported that they did not observe unethical behavior, while approximately 84% reported some unethical behavior.

In this study, the CVE model explained 31% of observed unethical behavior, which is a promising result. Six of the dimensions (white boxes in Figure 4.3) had a negative relationship with observed unethical behavior, in the following order of significance, as measured by standardized partial regression coefficients: (1) commitment to behave ethically, (2) ethical role modeling of supervisors, (3) ethical role modeling of management, (4) reinforcement of ethical behavior, (5) capability to behave ethically, and (6) openness to discuss ethical issues. But the other two dimensions (gray boxes in Figure 4.3) did not significantly support the respective hypotheses: clarity of ethical standards and visibility of (un)ethical behavior. Consequently, the existence of ethical standards (or codes of conduct) does not *necessarily* imply ethical behavior in organizations. This is not surprising.

Not surprising, as well, is the fact that the dimension "commitment to behave ethically" had the most significant negative relationship with unethical behavior. It also had the highest statistical confidence. Also, the "ethical role modeling of supervisors/management" showed notable significance. This supports the importance of "leading by example" that we discussed in Chapter 2.

Although there were some methodological shortcomings in Kaptein's pioneering study, it can be concluded that by using the CEV model to *expand* the concept of ethical culture, better management strategies could be developed to prevent, detect, and respond to unethical behavior in organizations [4]. It can be a useful tool for management as they develop and fine-tune their companies' approach to ethical business practices and define key areas of a value-driven culture of integrity. Naturally, this applies also to engineering organizations, bearing in mind that engineering (like every other profession) involves its own unique ethical issues.

In addition, organizations' business values frequently collide with engineering values. For instance, engineers aim to design products of high quality that last a long time and are maintenance-free. However, businesses may target products that are "good enough" to meet market demands, balancing cost and timely delivery, which can sometimes lead to compromises on certain aspects of quality and even ethics.

SUMMARY

Moral self-licensing and moral currencies—two central topics of this chapter—are not part of the standard content of engineering ethics textbooks. To our knowledge, the present textbook is the first of its kind to present tools from the field of behavioral ethics. These explanatory tools *complement* moral theories and related methods of ethical analysis, and facilitate the study of individual actors' viewpoints. The easy learning path of moral self-licensing involves the following discrete steps: (1) being aware of its existence; (2) understand what it is in detail; and (3) see that *conflicts* between granted licenses and one's conscience can arise—leading to an ethical hangover.

While moral self-licensing can be seen negatively as "fooling oneself" or "ethics-washing," we have to live with it because research on behavioral ethics has shown that we humans use this kind of procedure to give us permission to act unethically. Nevertheless, it is usually difficult to see and understand in advance all the relevant consequences of unethical actions because many of them are stochastic in nature, and the probabilities of their occurrence are low. This makes the "designing and tailoring" of moral self-licenses a "risky business."

If we do not need to go into the details of unethical actions in our explanations, moral currencies provide us with a tool that is particularly useful for gaining a high-level, general understanding of our own or others' unethical decisions and their typical give-and-take nature. This "giving and taking" is analogous to banking practices for private individuals.

And finally, in this behavioral ethics discussion, we moved from the individual level to the organizational level. There, an expanded view of ethical culture offers opportunities to manage the culture better and, as an outcome, reduce unethical behavior within and by organizations—in the long run. Furthermore, based on empirical research, it was shown that the multidimensional Corporate Ethical Virtues model has six (of eight) dimensions that are negatively related to perceived unethical behavior in real organizations. Hence, managing an ethical culture is a multifaceted endeavor for companies.

In conclusion, whenever contemplating different forms of unethical behavior to solve some problem, one should ask whether there are alternative ways of acting that would solve the problem at hand *without* unethical behavior. This is especially good to remember after previous discussions about moral self-licensing and moral currencies. In the following Chapters 5–7, we apply psychological and moral self-licensing to day-to-day cases in engineering organizations and one university involving *products*, *employment*, and *interaction*, respectively.

CLASS EXERCISES

1. At the beginning of this chapter is the text: "From the holistic perspective of engineering ethics, moral philosophy and behavioral ethics are complementary and not in competition with each other." Provide a justified explanation as to why this is true. Or do you think this is not the case?
2. Egon worked as a test engineer for embedded software in an elevator company. Due to heavy schedule pressure, he decided to omit approximately one-fourth of the standard test cases during the system testing phase. The earlier subsystem tests and the unit tests were successfully completed. Which of the virtues listed in Table 1.1 did Egon violate by behaving so?
3. Think of some accident from your past life. Was it a pure accident or actually a self-licensed outcome, if the *extended licensing* principle is applied? Justify your answer.
4. Does the *extended licensing* principle give an unreasonable interpretation of most accidents by classifying them as self-licensed—with only a few extreme exceptions? Remember that the sole purpose of this principle is to emphasize the role of individuals themselves in so-called accidents, and to prepare the reader to consider the following moral issues critically.

   Class debate: The instructor moderates a prepared debate between two motivated students. One of the students supports the negative answer ("it's unreasonable") and the other has the opposite opinion ("it's reasonable"). The rest of the class evaluates the arguments presented. What is the objective conclusion of the evaluation team?
5. Use one of the popular AI chatbots (ChatGPT, Copilot, or Gemini) to find possible weaknesses in moral self-licensing and in moral currencies.

   Class discussion: Compare the responses of different AI chatbots. Are there any significant differences between their answers?
6. On one Saturday afternoon, you lied to your good friend that you could not come to his/her birthday party because you were sick. Actually, you had

something "better" to do with another person. If your friend found out that you lied, what could be the consequences of your dishonesty? Use Table 4.3 for preparing your thoughtful answer.

7. Determine appropriate exchange rates (on a scale of 1–4) for the following virtues that could be used in applying the moral currencies system:
   - Fairness
   - Integrity
   - Respect
   - Responsibility
   - Tact

   Class discussion: Compare the answers of individual students and create a collective answer for the class.
8. Erich, a college sophomore, cheated on a written exam and was not caught. So, he took a moral-currency loan from his Ethics Bank. How, for instance, could he repay this loan to the bank? And why can he still get an ethical hangover even though he owes the bank nothing?
9. Kaptein observed in his empirical study that the existence of an ethics program in an organization, as such, "says very little about the frequency of observed unethical behavior in work groups" [4]. This result is somewhat surprising, because the purpose of ethics programs is, indeed, to improve the ethical culture of organizations. How does Kaptein explain this observation in his article?
10. Write a compact Ethics Pledge for students in this class. You can use an AI chatbot to assist you. In general, do such pledges have any real effect, or are they just empty words that are not taken seriously? If you utilized AI in the production of the pledge, what sort of impact does that fact have on your perception of the pledge's significance?

    Class discussion: Compare the pledges and answers of individual students, and create a collective pledge for the whole class.
11. Create a mind map of this chapter.

REFERENCES

1. M. Bazerman and F. Gino. "Behavioral ethics: Toward a deeper understanding of moral judgment and dishonesty," *Annual Review of Law and Social Science*, vol. 8, pp. 85–104, 2012, doi: 10.1146/annurev-lawsocsci-102811-173815
2. D. T. Miller and D. A. Effron, "Psychological license: When it is needed and how it functions," in *Advances in Experimental Social Psychology*, M. P. Zanna and J. M. Olson, Eds., San Diego, CA: Academic Press, 2010, vol. 43, ch. 3, pp. 115–155, doi: 10.1016/S0065-2601(10)43003-8
3. D. A. Effron, "Beyond 'being good frees us to be bad:' Moral self-licensing and the fabrication of moral credentials," in *Cheating, Corruption, and Concealment: Roots of Unethical Behavior*, J.-W. van Prooijen and P. A. M. van Lange, Eds., Cambridge, UK: Cambridge University Press, 2016, ch. 3, pp. 33–54, doi: 10.1017/CBO9781316225608
4. M. Kaptein, "Understanding unethical behavior by unraveling ethical culture," *Human Relations*, vol. 64, no. 6, pp. 843–869, 2011, doi: 10.1177/0018726710390536

5. D. M. Messick and A. E. Tenbrunsel, Eds., *Codes of Conduct: Behavioral Research into Business Ethics.* New York, NY: Russell Sage Foundation, 1996.
6. F. Gino and D. Ariely, "The dark side of creativity: Original thinkers can be more dishonest," *Journal of Personality and Social Psychology*, vol. 102, no. 3, pp. 445–459, 2013, doi: 10.1037/a0026406
7. M. Nicoll, *Dream Psychology.* First published in 1917. York Beach, ME: Samuel Weiser, Inc., 1987.
8. S. Bok, *Lying: Moral Choice in Public and Private Life.* New York, NY: Pantheon Books, 1978.
9. D. A. Effron and B. A. Helgason, "Moral inconsistency," in *Advances in Experimental Social Psychology*, B. Gawronski, Ed., San Diego, CA: Academic Press, 2023, vol. 67, ch. 1, pp. 1–72, doi: 10.1016/bs.aesp.2022.11.001
10. A. Giubilini (2023), "Conscience," in *The Stanford Encyclopedia of Philosophy*, E. N. Zalta and U. Nodelman, Eds. Accessed: Jan. 26, 2024. [Online]. Available: https://plato.stanford.edu/archives/fall2023/entries/conscience/
11. Z. W. Woessner, "Why good leaders choose to play the villain: The effects of moral licensing and perceived trust on leader behavior," M.A. Thesis, Department of Psychology, Michigan State University, East Lansing, MI, 2022. Accessed: Feb. 19, 2024. [Online]. Available: https://d.lib.msu.edu/etd/50817
12. D. Miller, "Beyond belief," book review of R. Smullyan, *Forever Undecided: A Puzzle Guide to Gödel.* New York, NY: Knopf, 1987, *Nature.* vol. 328, no. 6127, p. 212, 1987. Accessed: Mar. 29, 2024. [Online]. Available: https://www.nature.com/articles/328212b0.pdf
13. O. Weisel and S. Shalvi, "Moral currencies: Explaining corrupt collaboration," *Current Opinion in Psychology*, vol. 44, pp. 270–274, Apr. 2022, doi: 10.1016/j.copsyc.2021.08.034
14. A. Brei, "Being loyal and being ethical," *IEEE Potentials*, vol. 41, no. 3, pp. 46–48, 2022, doi: 10.1109/MPOT.2020.2989712
15. "Integrity Survey 2013." KPMG Forensic. Accessed: Feb. 19, 2024. [Online]. Available: https://assets.kpmg.com/content/dam/kpmg/pdf/2013/08/Integrity-Survey-2013-O-201307.pdf

5 Product-Related Micro-Level Cases

Learning Objectives

After studying Chapters 5–7, you will have the basic skills to

- Explain the decision-making process behind one's unethical behavior using moral self-licensing
- Analyze problematic behavior and actions in small, everyday cases using moral theories
- Evaluate the consequences of identified unethical actions for both individuals and organizations

In Chapters 5–7, carefully selected, detailed cases are presented and explained/analyzed with moral self-licensing and moral theoretical tools. All these micro-level studies have solid real-life roots, but the authors have adopted a certain authors' license to tell the stories, thereby preserving the anonymity of the individuals and organizations involved. Hence, we do not include the names of the organizations, just a number of companies of different sizes and representing different industries, from multiple continents, as well as one university. In addition, some of the relatively "thin" root cases were thoughtfully supplemented with relevant fictional specifics in order to enhance their pedagogical value.

Whenever something is built on authentic cases, one can ask—with tongue in cheek—whether they are models to imitate or cautionary examples to follow. But ours are neither; thoughtful explanations and comprehensive analyzes make them sound educational entities that promote the learning process toward ethical behavior among engineers and within engineering organizations.

In our demonstrative vignettes, we have aimed to reflect the gender distribution of current engineering students in the United States. Hence, the representation of female actors is 16.4% (9/55), which is close to the average percentage (16.3%) of female mechanical (17.9%) and electrical (14.6%) engineering graduates with bachelor's degrees in 2023 (*Source*: ira.asee.org/by-the-numbers/). And, there is a female actor in 45% (9/20) of these vignettes. Notably, the number of women in engineering is gradually increasing, which is gratifying to observe.

The pseudonyms used for the actors in the vignettes are Egon, Elke, Erich, Gunther, Heidi, Kurt, Walter, and Willi. At the end of each vignette, we assign a subjective actor's qualitative virtue measure (AQVM) to the relevant virtues of the actors. Our purpose is to stimulate discussions in the class about the cases

DOI: 10.1201/9781003485520-5

and their interpretations. This AQVM has three strength levels illustrated by emojis:

1. Remarkably virtuous
2. Fairly virtuous
3. Poorly virtuous

And the virtues considered in our cases are determination, fairness, honesty, integrity, loyalty, respect, responsibility, self-discipline, tact, tolerance, and truthfulness.

After each vignette, the main actions are given speculative moral (or psychological) self-licenses that illustrate the possible decision-making processes of the actors before they acted. This is the best we can do, because we cannot interview the anonymous actors themselves. References to specific licenses are denoted by the alphabet in curly brackets {A–Z}, in the case descriptions. Some of the licensees may not be considered morally relevant, but they offer a choice between virtues and vices or other problematic matters. A licensed actor here is a person whose restrictive psychological barrier has been removed [1]. Thus, he has successfully fabricated for himself a license to act—please remember that all the genuine actors have actually acted accordingly in their real-life settings.

Next, we analyze each vignette using a set of moral theoretical tools. Our goal is to answer the central question: "Were those self-licensed actions right or wrong?" This section seeks to assume the virtual role of the actor's conscience. While we do speculate about the actor's moral self-license [2], it would be pointless to speculate about the outcome of an individual's conscience [3]. On the other hand, moral theories give us a valuable insight into the question of right or wrong, which is reflected in the conscience of each actor, as well.

Each case study ends with a brief discussion of the vignette, and when applicable, provides potential consequences of its unethical actions: at the individual and local levels, as well as in the short and long term. It also includes, where relevant, suggestions on how that particular case would be handled in organizations with an advanced ethical culture. Thus, these micro-level case studies can be illustrated by the following process:

$$\text{CASE} \Rightarrow \text{AQVM} \Rightarrow \text{LICENSES} \Rightarrow \text{ANALYSIS} \Rightarrow \text{DISCUSSION}$$

Finally, after all case studies, we summarize the chapter with concluding remarks.

OVEROPTIMISM IN NEW-PRODUCT SALES

These two cases of the product category deal with bold risk-taking when launching sophisticated products that are still under development for sales. And in the first case, any delivery problems that might arise would probably be patched by bribery, which is one form of corruption.

We're Doing Business

BOX 5.1 VIGNETTE

ACTORS

Kurt: project manager
Walter: vice president of marketing

CASE

Kurt, an experienced systems engineer, was a project manager in a midsize HVAC (Heating, Ventilation, and Air Conditioning) company. His department developed HVAC automation systems for major office buildings and hotels, typically high-rise buildings. Kurt's ongoing project was designing a large-scale control and supervisory system. It was going to be their first product with web-based user interfaces for building supervision centers and with mobile apps for maintenance personnel.

Furthermore, this was their biggest and most challenging software development effort to date, and no more than 40% of the software existed already. Hence, it was only able to perform the core automation functions: sensor-data acquisition, communications, control, and status monitoring. But all the advanced operation modes, such as AI-based fault diagnostics, operational statistics, and system/node test modes, were still just specifications.

Walter, the vice president of marketing, had organized a task force to develop marketing materials for the upcoming product. Kurt's project engineers had helped them create different screen views that presented those versatile operation modes, but all the fancy screenshots were edited manually because the actual software did not exist. The online marketing materials produced were convincing; they gave a professional look to the intelligent HVAC control and supervisory system. And the new slogan for this flagship product was "Ultimate quality in HVAC automation."

Walter, a dominant extrovert, was getting old; he already seemed to be experiencing some sort of ego depletion. On one day, he told Kurt that they will soon be entering into a bidding process and offering the new automation system for a major HVAC modernization case in a 17-story office building {A}. Kurt exclaimed that it would be unwise to do so! Modernization cases have relatively short delivery times, and about 60% of the software—as well as all system tests—were still missing. Therefore, if they were to get this contract, the schedule pressure on Kurt's engineers would be unbearable. On the other hand, HVAC installations of new buildings would have much longer delivery times because those buildings must, naturally, be constructed first.

Walter grinned and said that you nerdy engineers do not understand business. If we run into trouble with their chief building manager, we can probably keep him happy by arranging a couple of "meeting weekends" with him in places like Las Vegas or Atlantic City—that usually helps {B}. Kurt kept trying to convince Walter that it was much too early to think about modernization cases, but Walter just said "we're doing business."

After a few weeks, Walter told Kurt enthusiastically that they had won the hard bidding competition, and the HVAC modernization contract had already been signed. Moreover, he stressed firmly that now is the time to do software development at full throttle! Kurt shook his head slowly—he did not trust Walter at all.

AQVM

Kurt: responsibility 🙂, loyalty 🙂
Walter: responsibility 🙁, respect 🙁, integrity 🙁

Licenses

{A} Walter's moral self-license consisted of three components: (1) based on his extensive experience in similar situations, he simply knew what should be done; (2) he wanted the new product to have as long a sales period as possible; and, most importantly, (3) he acted solely for the benefit of the company. This case shows the occasional attitudinal distance between marketing people and practicing engineers: risk-taking versus risk-bearing.

{B} Walter's moral self-license had four components: (1) a special "weekend meeting" in a place like Las Vegas was his well-proven recipe; (2) it would be a relatively inexpensive patching solution to potential schedule problems; (3) Walter had little respect for the chief building manager and was convinced he would take that bait; and again, (4) he would act for the benefit of the company. Apparently, Walter had licensed this kind of bribery into his toolbox already long ago, and thus did not see it as any problem.

Analysis

The most obvious moral issues in this case belong to Walter, so we will focus on those and leave it as an exercise for the reader to determine if Kurt is to blame (or praise) for any of his decisions and actions. We have just seen how Walter self-licensed his behavior, {A} and {B} above. Now, let us focus on how these conscious and subconscious acts of rationalizing fare when set against moral standards.

The Utilitarian Perspective

By entering the bidding process and agreeing to a delivery date, Walter was making a false promise. He was, in effect, telling the prospective customers that the HVAC automation system was further along in the development process than it actually was, and that it would be completed and in place by an unreasonable deadline. And even if "promise" is the wrong word for what marketing professionals like Walter offer to potential customers, he certainly agreed to terms that he was fairly confident could not be met. Call it what you will—deception, exaggeration, fabrication, or even bullshit [4]—a "false promise" is the phrase we use for the action Walter performed.

Now, to a Utilitarian, there is nothing wrong with false promises *per se*. In fact, misleading a potential customer might be considered the right thing to do, so long as the consequences of the act are sufficiently positive. Recall, a Utilitarian is one who believes that actions are right so long as they bring about the best overall balance of happiness over unhappiness for everyone affected. By this standard, it may seem as though Walter's false promise was exactly the right thing to do under the circumstances. With it, Walter secured the bid for his company, pleasing his marketing task force, his bosses, his new customer… and himself. With so much happiness going around, this false promise might appear justified. But recall Kurt's concerns and his reaction to hearing that his company had earned the bid. His disappointment over not having been listened to and his anxiety over the amount of work that lay ahead of him and his engineering team amount to displeasure and unhappiness. Therefore, Walter's action—the false promise—did not maximize the happiness of everyone affected by the action.

You might want to consider an alternate possibility. Had Walter *not* made a false promise, his company would likely not have gained a customer. That would have pleased Kurt, but would it have displeased Walter? So often in professional life—as well as life, generally—it is impossible to please everybody.

The Deontological Perspective

Now consider the situation from a Deontological perspective. What would Kant say about Walter's action? At first glance, it seems obvious: Kant says that lying is always wrong and Walter lied. Simple. But knowing the letter of the law is not the same (or as important) as understanding the spirit of the law… so let us understand why a deontologist like Kant would condemn Walter's actions.

Recall that Kant grounded his moral theory on just one moral principle: the Categorical Imperative. Also recall that there are at least a couple of ways to express that principle. According to one, the Principle of Universalizability (PU), morality requires us to only perform actions that we can *rationally will* everybody to perform. It is easy to say, "Sure, I would like it if everybody made false promises in order to win bids, and that's just what I'm going to do." But if false promises were morally required of everyone, no customer would ever believe any promise made during a bidding process. And if that were true, then Walter's goal of securing the bid would no longer be attainable in the way he intends. The clear misalignment between (a) our goals and (b) the actions we perform in order to achieve those goals is what indicates the wrongness of making false promises—to say nothing of bribery.

A deontologist might also express the Categorical Imperative in terms of the Principle of Humanity (PH), which instructs us to treat persons with respect, never as objects or tools to be used. Walter violated this principle by regarding the chief building manager as somebody who could be placated or "kept happy" with some weekend debauchery; written off as a business expense. Clearly, Walter does not hold the chief building manager in very high esteem. Neither does Walter appear to respect Kurt enough to acknowledge his concerns over schedule pressures and workload. Walter suggests that engineers—nerds, one and all—lack an understanding of how business is conducted. Condescension of this sort is rooted in disrespect.

This all probably seems awfully abstract. Kant is asking us to imagine a world in which everybody acts the same way (PU), one in which everyone possesses some invisible, hidden sort of value (PH). But we live in a world where there are deadlines and obstacles, where philosophical speculation can seem impractical and unnecessary. Walter had a job to do, and he did it. Probably by making a false promise, he did not intend for *everybody* to make false promises—only him. And perhaps he regards others as worthy of respect only when they cannot be used as tools for his own purposes.

But what makes Walter so special that the rules do not apply to him? Why should he get to play fast and loose with the truth while he expects everybody else to be honest? And why would he expect others to respect him when he only occasionally respects others? To regard yourself as the exception to the rule is the kind of irrational thinking that Kant objected to, and it is the source—from a deontological perspective—of Walter's wrongdoings.

The Virtue Theoretical Perspective

Finally, let us consider the situation from the perspective of Virtue Theory. It may seem natural to ask, "How would a virtue theorist evaluate Walter's actions?" But that is not exactly the right question to be asking from this perspective. As we explained in Chapter 3, actions are not the primary focus of a virtue theorist's attention. Rather, questions about *character* take priority. So, instead of asking if Walter's actions were right, we should ask if Walter was good. Put another way, we ought to ask ourselves whether or not Walter's actions are the sort that a virtuous person would perform.

Based on the details provided in the case above (Box 5.1), we can get a sense of the sort of person Walter is—the sort who makes false promises when it suits him, bribes other professionals to cover up negligence, disregards, and belittles his colleagues when they disagree with him, and values competition and winning above collaboration and restraint. For the virtue theorist, the problem is not just that Walter performed various actions—lying, bribing, belittling, and so on. The problem, primarily, is that these actions appear to be habitual. They seem to be a natural expression of the character that Walter has cultivated. In short, this is who Walter *is*.

"But this is how professional engineering *works*," you might say. "This is how business is *done;* it doesn't mean that's who Walter is all the time." And perhaps that is true. We can imagine that when he is with his friends and family, Walter is an honest person, a genuine listener, and a supportive partner. But to suggest that lying and bribing and disrespecting could be unacceptable in ordinary life but *acceptable* when one is "being a professional" is to grossly misunderstand what being a person is. Although the responsibilities and activities of a professional may be different from those in everyday life, the standards of morality are not.

Of course, there is no absolute rule against lying or bribing on a virtue theoretical view like Aristotle's. There is instead the insight that certain habits of character make us more or less likely to flourish as human beings. Thus, Aristotle would not instruct us to never lie—as Kant may. But he would point out that because Walter seems to have a habit of deceiving others, he has made it more difficult for others to get along with him. Aristotle would also point out that Walter's tendency to predict future success based on past successes is a lazy—and potentially dangerous—habit of character. When Walter says that weekends in Las Vegas or Atlantic City "usually help," he gives the impression that he would rather rely on tactics that have allowed him to dodge responsibility in the past than give each new situation the attention it deserves. This is not what a virtuous engineer does, much less a virtuous human.

So, it seems clear that despite Walter's moral self-licensing, his behavior does not stand up to moral scrutiny. His conscience may be clear, but he is fooling himself. He is one of those people who rarely suffers from an ethical hangover—at least in his professional life.

Discussion

The attitudinal distance between marketing people and engineers was clearly manifested in this case. Unfortunately, this kind of harmful distance is visible in too many companies, even though its existence benefits no one—especially companies.

It remains unknown whether Walter's company had an established or emerging ethical culture because Walter's deep-rooted unethical behavior was simply an inherent consequence of his character. And Walter, the vice president and "old dog," may find it difficult to change his behavior even after the company's possible ethics reform. As stated above, "this is who Walter is."

On the other hand, Kurt acted like a responsible and loyal project manager could be expected to act, but his point of view was ignored. Besides, it seems a little odd that Kurt's boss did not participate in those conversations between Walter and Kurt. His/her professional status could have helped convey Kurt's concerns to Walter.

An Uncomfortable Question

BOX 5.2 VIGNETTE

ACTORS

Erich: electrical engineer
Gunther: director of the customer company
Walter: boss, vice president of engineering

CASE

A young engineer, Erich, worked for a company in the field of industry applications. He graduated three years earlier with a degree in electrical engineering. Erich worked for a European corporation that transferred him to their U.S. subsidiary company for a couple of years. His responsibilities were related to technology transfer and software development in an established field of application. Moreover, he was the only person from the parent corporation who was working in the subsidiary premises.

Just before Erich's arrival in the United States, a pilot version of the company's new product was sold to a major customer in Salt Lake City, Utah.

During his first weeks with the subsidiary company, the product was installed and commissioned, and the customer was happy with the basic functionality of it. However, the advanced functions and features, which played a central role in securing the contract, were altogether missing. Therefore, after a few weeks of growing frustration, the customer became upset.

One day, a director of the customer company, Gunther, invited Erich to his office. The director saw Erich as a representative of the European corporation, not simply as a technical specialist, and that made the setting stressful for Erich. After a quick handshake and introduction, the director asked, "Why doesn't our new installation have those advanced functions and features that were agreed upon from the very beginning?"

Erich's brain began to work at a supercomputer's speed while he considered the feasible answers and their possible consequences. He rapidly identified two alternative answers:

1. (Truth) "Your installation does not have the advanced functions and features; the software to implement those does not yet exist."
2. (Untruth) "Our engineers in Europe are in the process of launching an upgraded version of those advanced functions and features, and the company decided to postpone this delivery for some time, because the new version will be more versatile and easier to use than the existing one."

Through a swift analysis, Erich reasoned that the first answer would definitely cause trouble for his company and likely also for himself. On the other hand, the second answer could protect the company and himself—at least for a while.

Erich was a true company man, so he chose the untruth—it took only a few seconds to make the decision {C}. Then there was a lengthy silence, while Gunther was considering that answer. Finally, he began to smile and said, "This sounds great; we could still wait for up to three months. Our showcase installation deserves the very latest and most advanced version." Gunther had swallowed Erich's lie.

After returning to his office, Erich told his boss, Walter, about the brief meeting with Gunther, and his untruthful answer to the uncomfortable question. Walter was grateful and praised Erich's loyalty and professionalism {D}. Fortunately, the delivery issue reached a happy end because the versatile new version became available in less than four months.

AQVM

Erich: honesty, respect, loyalty
Gunther: fairness, tact
Walter: respect, responsibility

Licenses

{C} Erich's moral self-license consisted of three parts: (1) he saw his company as being under an attack by the director of the customer company; (2) Erich knew (assumed) that the development of the missing functions and features would certainly be completed within a few months; and (3) he lied both for the benefit of his company and to get himself out of the uncomfortable situation. The root cause of this whole incident was that the new product was sold to the customer prematurely. And the secondary cause was that the director saw Erich as a representative of the European corporation, not just a technical specialist—and Erich accepted this questionable role without hesitation.

{D} Walter's moral self-license to praise Erich, even though he had lied, was simply: that altruistic lie was a service to the company, it benefitted them all. And the company was everything to Walter.

Analysis

Two issues stand out in this case: lying to a customer and prioritizing the interests of the company over any others. Those issues arise because of the decisions made (and actions taken) by Erich and Walter, and so, we will focus on them. The only question we have relating to Gunther involves his interaction with Erich. You might consider and discuss whether or not Gunther should have dealt more tactfully with that exchange, and whether or not that rises to the level of a moral issue. Our suggestion is that it probably does.

The Utilitarian Perspective

Because the utilitarian determines rightness and wrongness by looking at consequences alone, we should look at what resulted from Erich's lie. By lying about the advanced functions and features, Erich avoided an uncomfortable conversation with Gunther. Given the choices available to him in the moment, Erich did what made him the happiest. Gunther was pleased with Erich's promise. And Walter was pleased by having been spared the trouble of dealing with a disgruntled customer. It would seem that because telling a lie in this situation led to more happiness than the truth would have, Erich did the right thing.

One might object that Erich *did not* do the right thing at all—he did the expedient thing. That is, he simply took the easy way out. John Stuart Mill dealt with this sort of objection in Chapter II of his book *Utilitarianism*, explaining that sometimes the right course of action is the same as the expedient—convenient, quick, pragmatic—course of action [5]. But the expediency of an action is never what makes it right, says the utilitarian. Rather, what makes an action right is the production of happiness (or the prevention of unhappiness) that it brings about. On that account, sometimes the right action will also be the expedient action, as in Erich's case. Other times, the right thing to do and the expedient action will differ. It all depends on the consequences.

We ought to be wary of such a view, according to which the ends justify the means. The production of happiness is a fine thing, but what about when it involves deception and disrespect—or worse? Moreover, notice how much luck played a part in this case. What Erich promised Gunther turned out to be true. The new and improved software was being worked on and did become available in a reasonable amount of time. But when he made that promise, Erich had no way of knowing this. Although everything *seems* to have worked out for the best in this case, the rightness of Erich's action seems accidental. Morality demands better.

The Deontological Perspective

By now, we know how a deontologist would evaluate the issue of Erich lying to Gunther. Wanting to do well for one's company is just fine, but not if it requires one to lie. When you lie, you disrespect the person you lie to. You use them, you manipulate them. You devalue them by treating them in a way that they do not deserve to be treated—as an object. From this perspective, then, Erich did the wrong thing.

But the deontologist is not just concerned with actions. Motives are significant as well. It seems that Erich was motivated to lie by the desire to protect his interests and those of his company. He reasoned (correctly, we imagine) that if Gunther learned that Erich's company had secured a contract under false pretenses, there would be trouble. So, rather than deal with the harsh consequences, Erich decided to lie. In doing so, he allowed himself to be motivated by something other than a sense of duty—duty not just to his company, but to humanity in general. And so, from a deontological perspective, Erich is wrong not only because he performed a wrong action, but also because he was motivated in the wrong way.

The Virtue Theory Perspective

The item that stands out in this case is Walter's appraisal of Erich's lie as an act of loyalty. Recall, when Erich explains how his meeting with Gunther went, Walter expresses gratitude and praises Erich for his loyalty. Loyalty is a virtue, of course, and so ought to be praised. But *was* Erich loyal? A strong case could be made to support the claim that lying to a customer in order to cover up negligence and escape censure was not loyalty. After all, being loyal involves showing an appropriate amount of regard and respect to a person (group, creed, etc.). The key word here is "appropriate." Being loyal is not merely the act of showing respect—it is also the judgment that the person (company, ideology, or whatever) is *worthy* of that respect. Would we praise a Nazi officer for his loyalty to the Führer? Certainly not, because that loyalty is misplaced. In such a case, "fanaticism" is the proper term.

Extreme examples like this aside, the point is simply this: from a virtue ethical perspective, Erich was not wrong because he lied; he was wrong because he misjudged his company's worthiness. Had his company not misled its customer to begin with, perhaps Erich would have been right to protect its interests with a lie. But as things stand, "loyal" is not the correct word to describe Erich. Being a good engineer is not easy, and neither is being a virtuous engineer.

Discussion

The root cause of this incident was that Erich's company had sold the new product prematurely because the advanced functions and features were still under development. And this was not told to the customer because it would have had a negative effect on securing the contract. The product was supposed to be well-proven. Things like this happen all the time, "that's what business is all about."

Erich was a fairly inexperienced engineer, so he took on more responsibility for his company than he really needed to. Instead, he could have told Gunther that such delivery issues are beyond his responsibilities and that Gunther should contact Erich's boss Walter and request the schedule directly from him. In this way, Gunther would get the answer from an executive who both carries the responsibility of this delivery and also has the power to manage the corresponding development activities and personnel. Another, maybe more professional, response might be, "Unfortunately, I do not know the actual schedule for development of those features, but let me find out." And then, after discussing with Walter, Erich would get back to Gunther with a truthful answer.

From the moral theoretical analyses above, it is clear that Erich's behavior was ethically questionable. He lied to benefit himself and his company. But it probably was not an easy decision for him. Therefore, it is likely that Erich got an ethical hangover—not a severe one, but maybe a long-lasting one. On the other hand, Walter saw nothing wrong in his own behavior and would certainly do the same whenever needed. However, it is unwise to praise unethical behavior at workplace because it sends a mixed signal to employees. It would be better to have a constructive discussion about the whole case, instead.

Based on the vignette, it seems obvious that Erich's company did not have an ethics program, and that employees were only rewarded for results, not for the means of achieving them. Maybe the company should shape up, and begin to "do what's right."

SHORT-TERM SCHEDULE BENEFITS

The following three product cases discuss the consequences when tempting, short-term schedule benefits led to extra costs in the long term. They deal with diverse cases, where a company's product development (PD) instructions were bypassed; a self-confident engineer performed schedule management from his narrow perspective; and a relatively inexperienced engineer hesitated to say that he was unable to meet an intermediate deadline of his project.

Dying Batteries

BOX 5.3 VIGNETTE

ACTORS

Elke: technical sales manager
Walter: department manager
Willi: vice president of marketing

CASE

A relatively small Southern European elevator company had developed an energy-efficient drive system for low-rise elevators. This "Green Brake" utilized a battery bank to store the passenger elevator's braking energy and reuse it when needed. Furthermore, part of the stored energy was saved for passengers' rescue needs during possible power outages.

This novel product did not yet have direct competitors on the market. However, the product was not ready to be launched for sale. Their elevator factory had just a single prototype elevator in regular use. Only some dozens of passengers used it daily when entering or leaving the building, and it had worked as expected for a period of three months. Everyone was excited about this innovative drive system, but it would still take more than six months of documentation and extensive testing before it could be released to the market.

Vice president of marketing Willi and technical sales manager Elke put a lot of mental pressure on Walter, whose department had developed the new drive system. They wanted to start limited sales in just weeks {E}. Willi told that they could modernize one of the elevators of a famous fine-arts museum in a European capital; it would be an excellent sales reference. And eventually, their manipulative pressure exceeded Walter's tolerance. Thus, Walter reluctantly promised that his engineers and technicians would customize and build one drive system for this particular case {F}. However, he knew that such a decision violated their PD instructions, which he himself had approved.

Willi and Elke were excited, and the contract with the customer was signed within two weeks. After three months, the modernized elevator was already operational. There was even a display that showed the percentage of energy taken from the batteries instead of the power grid, over the past week. Everything seemed to work just fine.

However, after eight months of busy elevator traffic, seven days a week, this percentage was practically zero; the batteries were dead. Walter and his engineers were puzzled—what could be the cause? The dead battery bank was replaced swiftly by a new one.

And again, after about seven more months, the new batteries were dead. This time, Walter's engineers had already solved the annoying problem in the production version of that drive system. The battery technology was changed, the charger unit was redesigned, the drive module was modified, and this considerably repaired version was thoroughly type tested.

So, the elevator company decided to replace that faulty prototype drive system by the new production version—at its own expense. They clearly did not make any profit from this modernization case.

AQVM

Elke: responsibility ☹
Walter: tolerance ☹, responsibility ☹
Willi: responsibility ☹

Licenses

{E} There were three parts to Elke and Willi's psychological self-license: (1) a prestigious fine-arts museum would be a highly visible sales reference; (2) such opportunities were rarely available; and (3) all this was for the benefit of their company. Here, we do get a reminder that everything that seems to work is not necessarily ready for production.

{F} Walter's hesitant moral self-license consisted of four components: (1) so far, he only had positive and promising experiences with the new drive system; (2) he trusted his team's design expertise; (3) just this single special case; and (4) for the benefit of the company. This problematic case was launched when Walter, who had himself approved the PD instructions, allowed subjective external pressure to make him bypass his own instructions.

Analysis

The Utilitarian Perspective

A case like this is a good illustration of the significance of *interests* in ethical inquiry. In one way or another, utilitarians equate goodness with the satisfaction of interests. Elke and Willi were out to satisfy their interests in being good at their jobs, as well as the broader interests of their company. They might also have been considering the interests of the museum and its patrons. In their minds, securing a contract with the museum was a great way to do all that. But sometimes the satisfaction of one's interests comes at the cost of another's interests. In order to secure the contract, Elke and Willi had to pressure Walter to agree to an overambitious schedule, one that went against the company's standards for PD. It is reasonable to suggest that this went against Walter's interests, as well as those of the engineers and technicians he worked with. By committing them to a narrow timeframe, Elke and Willi did not appear to have the engineers' and technicians' interests in mind. Still, the elevator drive system was installed before the contract deadline—so, in the end, perhaps all is well from a utilitarian perspective.

But as we know, all was not well. The batteries could not keep up with the high elevator traffic, and had to be replaced twice—at no small cost to Walter's company, we must assume. So, were the interests of everybody involved actually satisfied? To answer that question, it helps to recognize the distinction between short-term interests and long-term interests. In the short term, a utilitarian might agree that everybody involved benefitted. Sales and Marketing secured the contract, engineering customized and built the drive system, the museum and its patrons got a new, "greener" elevator. But the consequences of our actions are not only experienced in the short term.

In the long term, it is hard to say that interests were satisfied for everyone involved. As a result of the pressure placed on Walter by Elke and Willi, the company had to (1) absorb costs that it would not have had to bear; and (2) do work that it would not have had to do, had PD gone according to normal standards. Is it reasonable to expect Elke, Willi, and Walter to see that far into the future? Because cutting corners and speeding up development timelines often lead to errors, it seems very reasonable. From a utilitarian perspective—and mindful of consequences in both the short- and long terms—Elke and Willi were wrong to push Walter as they did.

The Deontological Perspective

The issue that stands out in this case is the manipulating that went on. Walter was pressured to go against his better judgment by colleagues who seemed primarily concerned with their own goals. To a deontologist like Kant, this is a clear violation of the Principle of Humanity. According to that principle, we ought never to treat others as means to an end, as tools for our own use. Elke and Willi seemed perfectly happy to ignore Walter's advice and to push him to go along with the sale. Treating others this way amounts to disregarding their value. It cannot be justified in professional life or in life, generally.

But what would a deontologist say if this case had gone differently, and the elevator functioned perfectly? Would that justify the manipulation? Certainly not. A deontologist, recall, does not regard the consequences of an action as capable of making an action right or wrong. To treat another person as an object is never acceptable, regardless of the consequences.

It may be a worthwhile exercise for the reader to consider whether or not the manipulating in which Elke and Willi engaged violated the PU. That is, given the goal that Elke and Willi had (that is, making the sale) and given the fact that their pressure tactics helped them realize that goal, should we suppose that their action was something that everybody in their position ought to perform?

The Virtue Theory Perspective

The pressure that Elke and Willi placed on Walter indicates something of their characters. In this situation, they failed to take Walter's concerns as seriously as they ought to have. And they made working with them difficult for people like Walter. We probably all know somebody who we would prefer not to have on our team, somebody who is too assertive and does not listen to others. The proper functioning of a team (or company or society) relies on the integrity of its parts—and the proper functioning of a human being relies on the integrity of their character. Elke and Willi would do well to remember that patience is a virtue.

We might also question Walter's character here. Perhaps he was not as self-assured and courageous as he ought to have been in the face of pressure from his coworkers. Standing up to Elke and Willi may have been difficult, but the right thing is not always the easy thing. Of course, we have the benefit of hindsight regarding this case—something Walter did not have. Still, part of being a virtuous engineer involves knowing when to stand your ground and when to compromise. And when in doubt, err on the side of prudence and safety.

Discussion

Elke and Willi used manipulative pressure to get Walter to agree to their proposal to sell the unfinished product to a prestigious museum. This was wrong according to all three moral theories. Walter let their pressure to override the PD instructions he himself had approved. As the manager of the PD department, his responsibility was to ensure that all stages of the PD process were completed before any product was launched for sales. But Walter was not tough enough, and he was likely to be the one blamed when the problems came to light. Moreover, he may have got an ethical hangover, not the "hawks" Elke and Willi.

This is another case showing the attitudinal distance between marketing people and practicing engineers: risk-taking versus risk-bearing. And it is also a manifestation of the sick corporate culture. The company could consider starting an ethics program—with the commitment of senior management. A company-level code of conduct, which is trained for all personnel, would be needed. And Walter would certainly benefit from management training to develop his weak leadership skills.

A Matter of Honor

BOX 5.4 VIGNETTE

ACTORS

Gunther: project manager
Kurt: software engineer

CASE

Kurt was a skilled and self-motivated software engineer who paid attention to detail. He was assigned to a time-critical project that was going to develop a new control system for an industrial application in North America. It was a relatively small project, and Kurt had the full responsibility of its embedded software part. The necessary hardware existed already, but some electrical design was needed, too.

Gunther, an experienced engineer, was the part-time manager of that project in Europe. But the actual project was carried out in Canada because the first—and already sold—delivery was going to be in Toronto. The relatively tight time schedule made this project challenging for Kurt—he only had seven months to relocate to Canada and develop a rather complex piece of software.

At the beginning of the project, Gunther told Kurt that the software to be developed should utilize as much as possible the existing software of a similar application. This would mean that roughly two-thirds of the new software could be based on existing code components. By no means did Gunther want two parallel pieces of software implementing very similar functions because that would increase upgrade and maintenance costs in the long run.

Nevertheless, Kurt saw this unexpected constraint as a serious schedule risk because he did not have any prior knowledge of that existing software. In addition, he was aware that it was poorly documented. On the other hand, he knew his own capabilities well and could manage his time-usage effectively. Thus, he confidently decided to develop his software from scratch, but neither Gunther nor anyone else knew it {G}.

Kurt worked exceptionally hard during the seven months. This project was practically his whole life. There were obstacles, but Kurt overcame them all. And eight days before the control system was to be operational at the customer's plant, the software was completed. Kurt had finished his project on time; it was a matter of honor for him. Gunther was happy, Kurt's Canadian colleagues were happy, and the customer was happy. So, was that a true happy end to this story?

After a year, Kurt's company decided to start selling the new control system in Europe and Asia, as well. But those regions have their diverse safety regulations and other preferences. Kurt's software did not have a variety of such options available because it was designed strictly for the North American market. Finally, Gunther and the others realized that Kurt had actually developed his software all the way from scratch, instead of using the existing software components—as Gunther had instructed him to do.

If Kurt had utilized the existing software, there would have been no problems with European and Asian regulations and preferences because that software had already a full set of such options available. Eventually, it was decided to abandon Kurt's software and redesign the control system around the old software. Naturally, this took time and resources, but thus there was no need to rewrite any code for those many different options. Obviously, this was not a happy end for Kurt.

AQVM

Gunther (short term): responsibility
Kurt (short term): integrity, respect, responsibility
Gunther (long term): responsibility
Kurt (long term): integrity, respect, responsibility

Licenses

{G} Kurt's moral self-license had three components: (1) he recognized a serious schedule risk and wanted to eliminate it; (2) he had complete confidence in his own skills and time-management abilities; and (3) solely for the benefit of the company. Kurt was undoubtedly a company man, but still he caused harm and extra expenses for his company. So, why did he behave inconsistently? He focused blindly on his current project and its goals, without seeing or understanding the larger business context. It is not surprising that an internationally operating company launches a product found to be successful in one marketing area also to other markets, which may have a variety of additional requirements.

Analysis

The Utilitarian Perspective

As presented, this case asks exactly the right question: does this story end happily? In other words, did Kurt's decision to create the software from scratch bring about the most happiness (and/or least unhappiness) among all the people affected by his action? It is tempting to focus on the short-term consequences and say yes. After all, Kurt was pleased with his efforts, and so were his Canadian colleagues; Gunther was pleased that the project was completed well and on time; and the customer was pleased by the final product. It seems that happiness was maximized for everyone affected by Kurt's decision, making it the right one. And if Kurt's company had never attempted to sell the software in other parts of the world, that might be the case. But in the long-term, Kurt's decision made life more difficult for his colleagues and his company. It resulted in more problems and more work, and thus cannot be described by a utilitarian as the right course of action. This is an important lesson for an engineer who would determine right and wrong from this perspective: the relevant consequences are not limited to the immediate aftermath of a decision and action.

Another point is worth mentioning here. An engineer's job is not done at the final turn of a screw (or the final click of the mouse or the final stamp of a report). The consequences of one's actions extend outward like the ripples caused by dropping a pebble in a pond. And although a utilitarian probably would not hold a person accountable for consequences that happened years after an action, they certainly would not regard one's responsibility as "ending" with the immediate consequences of one's action.

The Deontological Perspective

Kurt appears to have been motivated by a sense of duty and a desire to do good work. For a deontologist, this is exactly the right way to be motivated. So, whether or not Kurt should be praised in this case comes down to the action he performed. Was his action the kind that everybody in a position like his ought to perform? If we were to focus on the hard work and initiative, we might be tempted to say that Kurt did the right thing by creating the software from scratch. But if we remind ourselves that this went against Gunther's instructions, then that temptation goes away. Should everyone always ignore their instructions, there would be no sense in issuing instructions. This illustrates one of the difficulties with the deontological perspective: from one angle the act is one of resourcefulness, while from another it is an act of defiance. Kurt's decision fails the deontological test because although he was motivated by honor and duty, he disregarded the instructions of his boss. Kurt thought he knew better—and he was wrong.

The Virtue Theory Perspective

As the title of this case suggests, Kurt regarded honor as something important. And he was correct. Now, be aware that honor means more than just being respected—it also means being *worthy* of that respect. In order to maintain his sense of honor, Kurt believed that he needed to work long hours and put forth a great deal of effort in order to deliver what he and his company promised their customer. Generally speaking, that seems perfectly reasonable. But when we recall how he went about that, we might wonder if Kurt went too far. After all, he decided not to learn about the existing software by reaching out to people knowledgeable about it. Instead, he decided to start fresh, to reinvent the wheel, as it were. Whatever Kurt possessed in terms of ingenuity and determination, he lacked in terms of foresight and regard.

Does this make Kurt a bad person? Probably not, assuming he does not make a habit of acting too extremely in the situations he faces. The important thing for a virtuous person—and, thus, a virtuous engineer—is to learn from their mistakes and make adjustments in the future. Integrity of character is a constant work in progress.

Discussion

Kurt appears to be an ambitious engineer, who was concerned about how others perceive him. And his desire for *honor*—and the admiration that goes with it—probably had a purely selfish basis, which could be related to low self-esteem. In this divided case, there are two dissimilar phases: the first *successful* phase and the second *not-so-successful* phase.

While "he did not have any prior knowledge of that existing software," he could have spent time with its developer(s), and learned the software architecture, real-time constraints, interfaces to the external environment, and so forth; "that's what engineers do all the time." Maybe Kurt was not a team-player, or he might have had some personal issues with the developer(s) of that software. Without such "excuses," it is hard to understand Kurt's decision, because "roughly two-thirds of the new software could be based on existing code components." Developing "two-thirds" of the software from scratch was certainly more time-consuming than learning the structure of existing software. A deontological viewpoint puts such behavior in a point-blank form, "he was wrong."

And the outcome of the second phase must have been a shock to Kurt: his behind-the-back actions became widely known, other engineers did the work he was expected to do a year ago, and he was no longer considered a hero. He felt he had lost his face—most likely, he got an ethical hangover.

This incident is strongly related to Kurt's character rather than the ethical culture of the company, but Gunther should have supervised Kurt's work and progress more carefully to make sure Kurt was actually following his instructions—"that's what project managers are for." On the other hand, Kurt used to be a trusted engineer.

A Mysterious Deadlock

BOX 5.5 VIGNETTE

ACTORS

Heidi: computer engineer
Kurt: leaving computer engineer

CASE

In his first project, Kurt, a newly graduated computer engineer, was developing a master–slave communications protocol for a distributed control application. Its design and programming phases were completed, and Kurt was finalizing the test phase. Almost all tests were passed, but then he noticed that the communications between one master unit and multiple slave units could sporadically enter into a locked state. This situation happened infrequently, once every few days. Hence, it was difficult to identify the reason for such a dramatic failure. For some time, it was assumed that the deadlock was hardware originated and caused by electromagnetic interferences. But no evidence to such a hypothetical problem was found. Kurt spent an intense couple of weeks trying to solve the problem; during that period, he was working 11 hours a day. However, he could not find the root reason behind the deadlock.

On the other hand, Kurt was about to leave the company for 12 months because he had to do his compulsory military service in his home country. Therefore, he wanted to leave his work desk empty, thus informing his project manager that the entire testing phase of the communications protocol was completed successfully. And he did not mention anything about the existence of the mysterious deadlock {H}.

Kurt's colleague and friend Heidi was going to continue Kurt's work with the ongoing development project, and she would also use the communications protocol Kurt had developed. Several months later, when Kurt was in the military, Heidi unexpectedly called him and asked about a rare deadlock condition that she had observed with the protocol. Corporal Kurt pretended to be surprised—"he had never noticed anything like that"—although the problem was well known to him {I}.

Heidi persistently continued to troubleshoot and eventually found the cause of the deadlock. The troublesome bug was fixed and the communications protocol worked perfectly after that. Moreover, Heidi did not find out that Kurt knew about the deadlock problem, but she just thought that Kurt did not run the system tests professionally enough.

AQVM

Heidi: respect 😐, responsibility 🙂
Kurt: honesty ☹️, respect ☹️, responsibility ☹️

Licenses

{H} Kurt's moral self-license had two components, and we speculate two options for the first one: (1a) soon he was going to act patriotically by entering military service—not every young man does that for various reasons; or (1b) his project manager put significant pressure on clearing the work desk before leaving; anyway, (2) he was going to fix the problem himself when he got back. This was Kurt's first project in industry and probably the missed deadline was the first he had ever missed as an engineer. So, he was ashamed and anxious, and thus overreacted with the issue in a questionable way. In fact, every engineer who has worked on PD projects for a while has also missed deadlines due to various uncertainties.

{I} This time, Kurt's moral self-license was obvious: he was reinforcing his earlier lie to keep it alive and to be consistent. As Sissela Bok wrote in her classic book, "it is easy, a wit observed, to tell a lie, but hard to tell only one" [6] (p. 25).

Analysis

The Utilitarian Perspective

Based on what we know from the case, it appears that very little went wrong here from a utilitarian point of view. Kurt lied more than once to his coworkers, it is true. But lying is only wrong, according to this view, when it fails to maximize happiness; and Kurt's lies do not appear to have caused any significant harm. Heidi was able to locate and fix the problem, and was never aware that she was solving a known problem. To put it simply, all is well that ends well.

But things could easily have gone a different way. What if Kurt's project manager or Heidi had discovered they had been lied to? What if problems with the communications protocol had caused significant and difficult issues? What if the lies he told ate away at Kurt's conscience, making him less effective in his military service? Naturally, one could ask a series of "what ifs" about any decision or action. What we are suggesting is that the fact that no serious harm resulted from Kurt's lies is something largely beyond Kurt's control. And to praise Kurt in this case is odd, since the rightness of his actions seems somewhat accidental. It seems clear that Kurt lied in order to protect his own interests, not to serve the greater good. The fact that everything worked out for the best is a fortunate accident, and morality demands more than that.

The Deontological Perspective

A deontologist like Kant would echo the sentiment just expressed—namely, that being a good person ought not to depend on any external factors. Your goodness depends on your motives and your actions—and that is all. You cannot control what other people do; neither can you control all of the circumstances in life that you face. All you can do is ensure that you do the right things for the right reason.

But should we suppose that Kurt was wrong to lie? From a deontological standpoint, almost certainly. By lying in order to satisfy his desire to tie up all loose ends and leave an empty work desk, Kurt acts irrationally. He fools himself if he believes that his goal can be reached in such a way. He also disrespects his coworkers, whom he regards as unworthy of the truth. So, while there may be nothing wrong with Kurt's desire to leave an empty work desk, he should not attempt to accomplish that goal by lying to cover up known flaws.

The Virtue Theory Perspective

There are at least a couple of ways to diagnose Kurt's dishonesty. One involves supposing that he intentionally lied out of a concern for his image. Because Kurt did not want to appear to be shirking responsibility, he lied to create the impression that he had worked diligently and effectively. To think along these lines is to suppose that Kurt's main concern is how other people see him, not with how he actually is. To a virtue theorist, that would be a problem. Being virtuous is not a matter of how others perceive you. It is about having the character traits that allow you to function properly as a human being. In short, it is about the inside, not the outside.

Perhaps a more charitable assessment is called for. We could understand Kurt's lie to have resulted from a sincere desire to avoid leaving a project unfinished. Perhaps he also wanted to avoid creating work for other people. If that were the case, then Kurt's lie is more of a miscalculation than a malicious deception. In this light, Kurt's sense of responsibility brought about an extreme act—something disproportionate to the situation. Of course, this does not absolve Kurt from blame. Despite his desire to act responsibly, he still acted extremely (i.e., viciously). More reflection and more practice are the remedy for that sort of moral misstep.

Discussion

Kurt was an inexperienced engineer who worked in his first development project. He apparently had unrealistic ideas about the project work and his own expertise: project's milestones *must* be met and he was already an *expert* in his field. And he tried to keep such an idealistic facade upright. But he did not succeed in completing his project task; maybe his expertise was not that great yet. This can happen to every engineer at the beginning of his or her career—such a fact is only to be accepted.

However, Kurt chose the irrational way of lying; he cheated both himself and the members of his project team. The milestones of the project cannot be met simply by hiding the truth. Kurt must have felt relief because he was not caught; he just "did not run the system tests professionally enough." On the other hand, this was also a negative outcome for Kurt because he might have learned (hopefully not!) that lying can solve schedule problems.

Kurt's unethical behavior had nothing to do with his company's ethical culture. He did what he wrongly thought was necessary. But to develop him into a *professional* engineer—referring to his overall development, not the professional engineer certification—he would benefit from training in project work and skills such as collaboration, time management, and critical thinking. Perhaps Kurt had not initially received proper orientation training for work, and he was put in the demanding project directly from the college bench.

PROBLEMATIC PATENTING PROCESSES

Intellectual property (IP) rights are an integral part of innovation and PD in engineering organizations. Patents are an important class of IP. The following product cases deal with specific problems that arose during the patenting processes of two electronics companies. In the first case, the IP rights of a consulting engineer were unjustly withheld behind his back. And in the other case, a distinguished scientist had an inexplicable attitude problem while preparing an expert statement for the European Patent Office (EPO).

Consultant's Invention

BOX 5.6 VIGNETTE

ACTORS

Egon: boss, vice president of engineering
Erich: team leader
Gunther: consulting engineer

CASE

Erich's electronic systems team was developing a large programmable display panel for shopping malls and other public buildings. It was targeted for the Southeast Asian market. Therefore, it supported Chinese and other non-alphabetical characters. That made the design task somewhat challenging.

Gunther, an independent consulting engineer, was hired to design a handheld device for programming such displays easily on site. He had experience with similar devices. During the project, Gunther created a straightforward procedure and an associated user interface, with which non-alphabetic characters were conveniently entered into the programming device. Erich's boss Egon thought that the novel procedure should definitely be patented, and he asked Erich to start the patent application process immediately.

Egon also said that since Gunther was a consultant and not "our boy," he should not be named as the inventor, but Erich should be the sole inventor. Fortunately (from Egon's point of view), issues regarding IP rights were not addressed in their contract at all. Erich opposed strongly: "But this is purely Gunther's invention, not mine; such an arrangement is fair neither to him nor to me." "We are already paying a good amount of money for Gunther's consulting work," Egon continued, "I don't want to pay him any extra patent rewards or anything like that" {J}. Erich was confused and tried to resist, but Egon had made his decision.

Eventually, Erich prepared the patent application with a patent engineer of his company, according to Egon's instructions {K}. And nearly two years later, the patent was finally granted. So, Erich held now a new patent, although he was not its actual inventor; he was embarrassed and never included it on his CV. In addition, Erich received a standard $1,800 patent reward from the company—he felt guilty.

Gunther did not find out that his invention was fraudulently patented behind his back, and his consulting work with Erich's company continued successfully for many years.

AQVM

Egon: integrity ☹, respect ☹, responsibility ☹
Erich: truthfulness ☹, respect 😐
Gunther: respect 🙂

Licenses

{J} Egon's moral self-license consisted of three components: (1) he had no respect for well-paid consultants as individuals; (2) consultants were just his servants, whom he could use like "things;" and (3) solely for the benefit of his company. Such a disrespectful attitude is certainly not sustainable. Maybe there was some personal issue between Egon and Gunther.

{K} Erich's hesitant moral self-license had also three parts: (1) it really was not his idea—he was practically forced to do it; (2) he wanted to please his strong-willed boss; and (3) for both the company and himself. The organizational culture was sick.

Analysis

The Utilitarian Perspective

Rights are a strange subject for the utilitarian. Generally speaking, rights are regarded as legitimate claims either for or against certain kinds of treatments [7]. Erich believed that Gunther had a legitimate claim to be recognized as the inventor, a claim that would involve, among other things, financial benefits. Egon obviously believed otherwise. But what sort of moral foundation for rights can be provided by a utilitarian? According to this theory, the only legitimate claim that anybody has is to be considered in any assessment of the consequences of an action—when that action would affect the person in question, that is. Put more plainly, if you perform an action that affects my happiness (or unhappiness), then my happiness (or unhappiness) deserves to be considered. One might say I have a right to be part of the evaluation of your action, insofar as I was affected.

But according to the case, Gunther never found out about the patent. And if he was made neither happy nor unhappy by the act of patenting the invention, then there is nothing to factor in on his part. Remember, the only thing that a utilitarian regards as intrinsically valuable is happiness (a.k.a. pleasure). People are of secondary importance, only indirectly valuable as potential "experiencers" of happiness. To build a robust account of rights on a foundation like that is tricky.

In the final analysis, a utilitarian would probably praise Egon for his decision to leave Gunther out of the patent process. Sure, Erich experienced some guilt. But that is overshadowed by the fact that Egon met the customers' requests, kept the rewards of the patent "in house," *and* ensured a working relationship with the consulting firm. Had Egon done anything else—like been up-front with Gunther—happiness might not have been maximized.

The Deontological Perspective

The above account of Egon's moral self-licensing pinpoints the wrongness of his action. Egon fails to respect Gunther by not giving him credit for the invention. He also shows very little respect for consulting engineers or, indeed, anybody not directly employed by his company. Respecting others involves, among other things, giving them what they are due. In general, that means treating others in ways that are consistent with the *intrinsic value* they possess—that is, the value they possess in and of themselves, by virtue of being human. More specifically to this case, respecting Gunther would involve recognizing him for the good work he has done and wanting him to receive credit for it. Egon certainly fails to do that.

Moreover, the tribalism that Egon displays is worrisome. By referring to Gunther as "not our boy," Egon draws an arbitrary line between those who matter morally and those who do not. He appears to think of being employed by his company as necessary for full moral consideration, while *not* being employed by his company warrants disregard. One might notice similar instances of preferentialism relating to nationality or a sports team. No matter the context, this sort of arbitrary valuing is irrational, and it cannot justify Egon's decision.

Neither does Egon seem to respect Erich. Although Erich protests, Egon does not listen. Egon appears to have made his mind up about where the credit for the invention ought to go, and does not consider Erich's reasons worth listening to. Who would want to work for someone like that?

The Virtue Theory Perspective

One thing that stands out in this case is Erich's reaction to Egon's orders. When he was told to take credit for Gunther's invention, Erich was confused and resistant. These are signs that Erich has cultivated a certain set of virtues that disincline him toward deception and hubris. Of course, possessing virtues does not guarantee that a person will never act wrongly—as Erich does in this case. But a virtue theorist recognizes that there will be wrong turns and detours on the road to virtue. We might forgive Erich this one transgression, but we also might worry that giving in to orders that he knows are wrong might become habitual. And with a boss like Egon, the chances of such a thing developing are fairly high. An engineer like Erich needs to take special care that he does not compromise his values so much that he becomes inclined toward vicious behavior. He needs to stand up to bosses like Egon, even if doing so puts him at risk of being fired. That is easy to say and hard to do, we know. But nobody ever said that being virtuous would be easy—only authentic and good.

One other thing stands out in this case. The issue of IP was left out of the contract. And although we do not know for sure based on the details provided in the case, it is not outrageous to assume that Egon wanted it that way. He certainly did nothing to amend the contract in light of new developments. This eagerness to act solely on the letter of the law, as it were, instead of the spirit indicates a defect in Egon's character. A person who is inclined to hide behind a legal document like a contract in order to avoid responsibility reveals either an inflated sense of self-importance (pridefulness) or an incomplete understanding of their responsibilities toward others—or both. The standards set by a contract are not the same as those set by morality. To mistake the one for the other is a character flaw.

Discussion

The above deontological analysis ends with the words: "Who would want to work for someone like that?" This is a good question because it is very likely that this case was not the only time Egon behaved irrationally.

Let us open the curtain that protects the genuine company a little. It was a fairly large company with established procedures for dealing with IP rights with consultants, joint ventures, and universities. And Egon followed these principles routinely without any problems. Therefore, his strange behavior in this particular case must have been related to some personal issue between Egon and Gunther. At least, their personalities were very different—Gunther was a bit of a bohemian, which might have annoyed Egon.

Although Egon was the vice president of engineering and was apparently considered a competent leader, his ability to work with different types of people was not sufficient. In addition, he was vain and his leadership style was coercive. Maybe these negative qualities explain Egon's behavior. But since Egon was a true *company man*, a company-wide ethics program could open his eyes and he would adjust his behavior as needed, "for the good of my company."

Without considerable adjustments, Egon's behavior would undermine the work morale of his subordinates, including Erich, who appeared to be in a subjugated position. Egon did *lead by example*, but his example was wrong. Now, we return to the question at the beginning of this discussion, "Who would want to work for someone like that?" And our tongue-in-cheek answer is that "we don't know."

Expert's Statement

BOX 5.7 VIGNETTE

ACTORS

Erich: electrical engineer
Willi: distinguished scientist

CASE

Erich was working in a small project that developed an advanced sensor system for border security applications. He had created a novel procedure for intrusion detection in harsh environments using multiple microwave radars and digital signal processing. A patent application was prepared and filed to the EPO. Erich was named the sole inventor, and he was proud of his invention and this first patent application.

After a few months, Erich received a letter from the EPO. It was from a patent examiner who had studied the patent description and wanted some additional information and clarifications. Erich spent several days for preparing his thorough response to the patent examiner. After such a careful research and writing effort, he was convinced that all concerns of the patent examiner were satisfied.

But after two months, he received another letter from the same examiner—he became excited. This time, the patent examiner suspected the theoretical foundation of the invention, and was ready to reject Erich's application if he did not receive convincing explanations of its theoretical basis. Erich did not have anything to add to his previous response. Therefore, a distinguished scientist, Willi, was invited to write an "expert statement" to the EPO.

Willi wrote a lengthy statement where he explained rigorously that there are no theoretical obstacles why this invention would not work as described in the application text. And at the end, Willi wrote that he had personally seen the sensor system in operation, and it functioned exactly as described in the patent application text. But Willi had never seen the early (prototype) system in operation.

Erich was confused of that last sentence; how could Willi say in his official statement something that is not true? Erich told Willi that the last sentence is untrue. Willi just grinned and replied that the patent officer does not know it {L}. After a short negotiation, Erich suggested to Willi that the untrue sentence would be removed from the statement. Finally, Willi shrugged and agreed, and his expert statement was mailed to the EPO.

In a few weeks, Erich received a notification that his patent application was approved. He was relieved and delighted of the outcome. But still, he could not understand why such a distinguished scientist was going to lie in his official statement to the patent authority.

AQVM

Erich: integrity 🙂, respect 🙂

Willi: integrity 🙁, respect 🙁, responsibility 🙁

Licenses

{L} Willi's moral self-license was based on three elements: (1) he saw himself as an admired scientist in the field of radar signal processing; (2) he was confident that the sensor system would work as described; and (3) he wanted to make his strong statement even more convincing. Yet, perhaps, the ultimate reason for Willi's lying was his big ego.

Analysis

The Utilitarian Perspective

If the consequences alone determine whether an action was right or wrong, then it would be hard to identify any wrongdoing in this case. Perhaps the place to focus is on Erich's confusion over Willi's willingness to lie in an official statement. This particular act of lying resulted in a negative response on Erich's part. He must have been made to feel uncomfortable, which amounts to displeasure from a utilitarian perspective. Of course, Willi relented and removed the untrue statement. But the damage had been done—Willi failed to maximize the happiness of those affected by his actions.

This may be a good opportunity to step back and take note of something important. Based on the case analyses you have read so far, you may have formed the impression that the only job of a moral theory is to identify wrong actions—or, in the case of Virtue Theory, bad characters. Because we have only highlighted how the players in these cases fail to meet moral standards, that might be an understandable impression. But know that moral theories are just as capable of identifying and explaining wrongness as they are rightness. In other words, moral theories are meant to provide guidance for making right decisions and developing good traits. In that spirit, a utilitarian would likely praise Willi for agreeing to remove the lie from his statement, and Erich for insisting on it. But we digress.

The Deontological Perspective

Wanting to help Erich by writing a strong statement of support is admirable on Willi's part. But to end it with a lie surely goes against deontological standards. As we know, lying is seen as an irrational act of disrespect by a deontologist. We also know that the deontologist regards two elements of an act as jointly determining its rightness: the act itself *and* the motivation behind it. Thus, Willi may have had the right motives, but he relied on the wrong action.

For his part, Erich displayed good sense in this case. Although he wanted his patent application to be approved, he appeared unwilling to resort to lies to achieve that end. That is praiseworthy from the point of view of a moral theory according to which ends *do not* justify means.

The Virtue Theory Perspective

Based on the account above of his moral self-licensing, Willi has a high impression of himself. That is not entirely undue, of course. He is a highly-accomplished scientist. And assuming that he is not fooling himself, he is also highly regarded by the members of the scientific community. Taking pride in one's accomplishments is not a bad thing; neither is feeling good about the respect and acclaim one receives (as long as one is deserving). But pride is a virtue, and that means it lies on a spectrum between two extremes. In the case of pride, those extremes involve (a) having too low an opinion of oneself and (b) having too high an opinion of oneself. People with too low an opinion of themselves are prone to self-deprecation, to downplaying their value and accomplishments. Those with too high an opinion of themselves (like Willi) tend to overestimate their value and think of themselves and their accomplishments as greater than they actually are. Such a person—and we all know someone like this—tends to be difficult to work with and to be around, and is therefore not excelling in this dimension of being human. One can live with vices, but not live well.

Discussion

From the point of view of morality, this case had a happy ending. Eventually, nothing wrong happened. And maybe Erich's company had an ethics program going on, because he was so strongly against lying. Or perhaps he was one of the good people who have naturally high morale—even without special training or ethics programs.

However, it seems that Erich was pretty upset when Willi included a blatant lie in his official statement to the EPO. Erich had thought that Willi was not only a brilliant scientist, but also a virtuous person; a kind of role model for many. This case may have stirred up conflicting thoughts in Erich's mind: if it was so easy to write a lie into such a statement—because "the patent officer does not know"—could there also be deception behind Willi's admired accomplishments? Huh, we hope not! This unfortunate incident lowered Erich's respect and trust in Willi, his longtime idol.

EXAGGERATION WHEN PURCHASING EQUIPMENT

It is a stereotype that salespersons sometimes use questionable practices when selling equipment to engineering organizations. However, there is also a fuzzy group of purchasers with an engineering background who sometimes use similar means during bargaining processes. The last case in the product category is an example of the computer purchasing process where a tough purchaser crossed the line of dishonesty.

LINUX WORKSTATIONS

BOX 5.8 VIGNETTE

ACTORS

Erich: group leader
Heidi: sales representative of vendor-A
Walter: boss, department manager
Willi: sales representative of vendor-B

CASE

Erich, an experienced computer engineer, was hired as a group leader to a large multinational electronics and automation firm. His first task was to found a new microelectronics design group in Denmark. Among other things, he was going to purchase several Linux workstations for the newly hired engineers of the center. Walter, Erich's boss, advised him to begin a bidding process with workstation vendors A and B, because these vendors represented two competitive workstation brands.

Erich contacted Heidi (vendor-A) and Willi (vendor-B), and requested them to offer nine Linux workstations, six low-end versions and three high-end units with powerful 12-core CPUs, to be delivered to his design group by the end of the year. In addition, he said that they would likely purchase 60–70 workstations in the coming two years. His design group would grow fast and substantially, and 20–30% of the future workstations would be for high-end number crunching. However, this estimate was just Erich's (exaggerated) personal guess—it was not based on any company-level plans or even preliminary discussions. Erich just wanted to show his new boss how good prices he can get.

Heidi and Willi saw this bidding as an opportunity to open the door for future sales of larger quantities to this respected company. Thus, they offered their workstations at very competitive prices. But Erich wanted more. He negotiated toughly with Heidi and Willi, and was able to push the initial prices downwards. There was fierce competition between vendors A and B. Finally, Erich lied to Heidi that Willi had offered a certain total price, and if Heidi could go below it, she would get the deal; and he also lied in a similar way to Willi {M}. Hence, Heidi and Willi gave their very final bids; these were some 20% below their initial bids—Erich's bargaining had led to an excellent outcome. And Walter praised Erich as a skilled purchaser.

Willi's bid was slightly lower and he won the deal. Willi was exhausted, he had never experienced such hard bargaining before. But the future was not as rosy as Erich had envisioned for Heidi and Willi. During the next two years, only five more workstations were purchased for the design group, and none of those were the high-end ones. Erich's company had decided that all number-crunching computations, such as major circuit-level simulations, would be run on cloud servers in commercial data centers, instead of running on their own high-end workstations. Willi felt disappointed.

After some time, Erich switched to another electronics design company in Sweden. And he was going to purchase some Linux workstations for his new microelectronics team. Therefore, he met Willi again. When Erich and Willi began to negotiate about a possible bidding process, Willi said bluntly that he did not want to deal with Erich anymore—last time he had given such low prices that he did not get any sales bonus at all {N}. Afterwards, he had realized that Erich's future forecast had been just bullshit.

AQVM

Erich: honesty, respect
Heidi: respect
Walter: respect
Willi: respect

Licenses

{M} Erich's moral self-license had four parts: (1) my company is mighty; (2) those vendors are dependent on their clients; (3) they are lower in the professional hierarchy; and (4) for our benefit. This seems to be more of an ego issue than an explicit pursuit of company benefit.

{N} Willi's psychological self-license contained three components: (1) he had lost his trust in Erich; (2) his pride prevented him from dealing with Erich anymore; and (3) he had a right to choose his customers. Nonetheless, it is likely that a colleague of Willi represented vendor-B in Erich's new bidding competition.

Analysis

The Utilitarian Perspective

The bidding process described above is probably not unlike many similar situations in the engineering profession. The pressure is high and the competition is fierce. In circumstances like that, it is not surprising that questionable tactics would be employed. In this particular case, Erich lied to Heidi and Willi in order to persuade them to lower their prices. "That's how the game is played," we can imagine someone saying. And perhaps that is true. But that has got nothing to do with how the "game" *ought* to be played.

The initial consequences of Erich's lie include displeasing Heidi. They also include the happiness experienced by Erich himself and, we may suppose, his work group and his supervisors. Willi, it might be assumed, was pleased to win the bid. But the process exhausted him and left him without a sales bonus, so it is debatable how much happiness he experienced. Of course, the consequences of an action follow into the longer term, where we note that Willi realizes that he had been lied to. Accordingly, it would be difficult to imagine a utilitarian making an overall positive assessment of this case. Erich seems to have cared more about his own interests (and the consequences affecting them) than he did about the wider consequences. That is more self-interest than a utilitarian could abide.

The Deontological Perspective

This case illustrates beautifully the kind of irrationality that a deontologist like Kant finds so objectionable. Consider Erich's goal: to receive the lowest price possible for the workstations. Consider the action he performs in pursuit of that goal: he lies to Heidi and Willi. You might think, "It may be irrational, but it worked!" But recall what Kant's PU tells us. We ought only to perform actions that would allow us to accomplish our goals, even in a world where everybody else acted similarly.

What would result from everybody lying to secure bids? Well, at first, vendors might be fooled (as Willi was). But it would not take long before vendors stopped believing people like Erich and stopped offering low prices in exchange for (promised) high volume (as Willi did). The upshot is that in a world like this, Erich's goal of receiving low prices *could not* be achieved by making a false promise. And in that world, saying, "I am going to get low prices by lying to vendors" would be just as irrational as saying, "I am going to boil this water by heating it to 65°C" (by the way, the boiling point of water even at the top of Mount Everest is about 69°C). The thought experiment we are invited to conduct by the PU shows clearly how Erich fails to meet the moral standard set by the deontologist.

The Virtue Theory Perspective

Let us focus on the sentiment that is expressed in both of the above perspectives: the notion that lies are "part of the game" because they "work." We saw how a utilitarian and a deontologist would respond to this claim. What about a virtue theorist?

Perhaps the most obvious thing to say from this perspective is that you are only responsible for the moral character of one person: *yourself.* How others behave may present you with temptations and constrain your opportunities for practicing virtuous behavior. Aristotle was very forthright in acknowledging this. He knew that different people face different challenges and have different opportunities, talents, and connections. None of that, he claimed, changes what it means to be a good person. All it means is that developing virtuous habits will be more difficult for some people than it will be for others. That is life. Each of us is still responsible for who we are, and the circumstances we find ourselves in cannot be blamed for our failures of character.

We object to the suggestion that professional life amounts to no more than a "game." It is far too significant to be trivialized like that, certainly in the case of engineering. But our objection having been registered, we can say this: even if one's profession *were* just a game, that would still mean that it was governed by rules. And when others break those rules, it does not mean you are a good person when you break them too. Thus, Erich cannot be absolved for his dishonesty and disrespect.

Discussion

If this case were presented to the engineers who purchase equipment for their companies, many might just say, "Oh well, that's how the game is played." But, remember, an engineer's career is definitely not a game, but a serious undertaking. From the case story, it can be interpreted that Erich's main purpose was not to save his company's money when he purchased those workstations; the main point was to show how skillfully Erich could play his "game." He probably did it because it made him feel important and also showed Heidi and Willi how important he was; that is called ego-tripping. But a *purely altruistic* motive would have made Erich's behavior easier to sympathize.

On the other hand, Willi's behavior was somewhat surprising when he said that "he did not want to deal with Erich anymore." Generally speaking, salespeople have such thick skin that they will not ignore a single potential customer, no matter how nasty the customer may be. It is a part of their professionalism.

SUMMARY

This chapter is the first part of our triad of micro-level case studies. All eight discussed cases are related to products through everyday examples of PD, patenting of innovations, and the acquisition of workstations.

For each case, we first created speculative moral self-licenses to gain insight into the absorbing question, *why* was the particular action taken? Although these licenses are based on our intelligent imagination, built on decades of work experience, we believe that they provide a sound platform for our case studies. It is important that readers spend some time thinking that those unethical actions not just happened, but were licensed by the actor himself. Thus, in a way, the actor had permission to behave unethically.

Table 5.1 shows the main reasons for acting unethically in these product-related cases. "Company's benefit" was the primary reason in almost half of the licenses, "Actor's benefit" in a third of the licenses, while a fifth fall under the category "Other" (miscellaneous). In fact, it is well known that altruistic unethical behavior on behalf of one's company is fairly easy to sympathize.

Next, the licensed action was taken, and it would be time for the actor's conscience to analyze its details, and make a classification between *tolerable* and *intolerable* (see Figure 2.1). We used three moral theories as a "virtual" conscience to understand how serious the action was from the Utilitarian, Deontological, and Virtue Theory perspective. Based on our analyses, we found that the moral judgments of these different perspectives varied considerably in some cases. This observation correlates well with the difficulty we can encounter in justifying the seriousness of an

TABLE 5.1
Main Reasons behind the Speculated Psychological/Moral Self-Licenses {A} to {N}

Self-license	Company's Benefit	Actor's Benefit	Other
A	Yes		
B	Yes		
C	Yes	Yes	
D	Yes		
E	Yes		
F	Yes		
G	Yes		
H		Yes	
I		Yes	
J			Yes
K		Yes	
L			Yes
M			Yes
N		Yes	
%	**47**	**33**	**20**

unethical action—very often, our pragmatic justification is semi-objective, at best. In addition, we cannot reliably assess which of those unethical actors suffered from an ethical hangover. Only Erich in the vignette "An Uncomfortable Question" (Box 5.2) and Kurt in the vignette "A Matter of Honor" (Box 5.4), almost certainly got an ethical hangover after their questionable actions—and perhaps also Walter in the vignette "Dying Batteries" (Box 5.3).

From Table 5.2 we can see, for example, which of the identified virtues were the most violated. If we combine honesty, integrity, and truthfulness into a single meta-virtue "HIT," we can see that in addition to this *HIT*, *respect* and *responsibility* are

TABLE 5.2
Actor's Qualitative Virtue Measures in Product-Related Cases

Virtue	Strength Level	Strength Level	Strength Level
Fairness	0	1	0
Honesty	0	0	**3**
Integrity	1	1	**4**
Loyalty	2	0	0
Respect	3	5	8
Responsibility	4	1	9
Tact	0	1	0
Tolerance	0	0	1
Truthfulness	0	0	**1**

the other dominant virtues that were seriously violated. And all of them are particularly important for the working atmosphere. Furthermore, there is a connection between these three virtues, for instance, if HIT is violated, it means that respect is also violated—but this is not necessarily true the other way around.

To make a simple recommendation, we could say "let's focus on keeping the HIT violations to a minimum." Of course, this would not exterminate the problem of unethical behavior, but it would be the right direction to seek relief—as "honesty is the best policy." Remember that violating the meta-virtue of HIT is a key reason for reducing *trust* between individuals, or between individuals and companies. Moreover, in most cases analyzed, it was clear that the company did not have an ethics program. This is certainly an important issue to take into account. Fortunately, more and more engineering companies are already navigating toward an ethical pathway. Besides, a good engineer will always place safety of the public as the highest value in their decision-making, and exaggeration is a character flaw that plays no role in the professional character of an engineer.

The following chapter is similar to this one, but there the unethical cases are related to employment issues in engineering organizations. Common topics covered include: discrimination, layoffs, and business trips.

CLASS EXERCISES

1. Using the following moral theoretical perspectives, determine if Kurt should be blamed or praised for his decisions and actions in the vignette "We're Doing Business" (Box 5.1).
   a. Utilitarian perspective
   b. Deontological perspective
   c. Virtue theory perspective
2. Form an acting group of three volunteer students. The group first prepares a script for a role-play based on the vignette "Dying Batteries" (Box 5.3). And then they perform it in front of the class.

   Class discussion: What new issues related to ethics in engineering did the role-play raise compared to the text of the book alone? The actors' viewpoint versus the audience's viewpoint.
3. Did the manipulating in which Elke and Willi engaged in Box 5.3 violate the Principle of Universalizability (Chapter 3)?
4. Edit the case story of the vignette "A Mysterious Deadlock" (Box 5.5) into an ethical one. Would Kurt have lost anything if he had acted according to the ethicalized story in the first place? Or was lying his only valid option?
5. At the end of the vignette, "Linux Workstations" (Box 5.8), are the AQVMs given below.

   Erich: honesty 🙁, respect 🙁
   Heidi: respect 😐
   Walter: respect 🙂
   Willi: respect 🙁

   Explain the authors' possible thinking behind these selected virtues and their strength levels. Is there something that you would disagree with?

6. Table 5.1 summarizes the main reasons behind the speculated psychological/moral self-licenses; and nearly half (47%) of them are for "Company's benefit."

   Class discussion: Is it more acceptable to behave unethically for the benefit of your company than for yourself? Is it less likely to get an ethical hangover if the actor feels that he himself did not benefit from the unethical action at all? Can it be considered loyalty if someone behaves unethically to benefit his company? Is there some demand/pressure to act unethically to benefit one's company?
7. Which one of the above cases (Box 5.1–Box 5.8) was the most against your own ethical principles? Justify your answer.

   Class discussion: Compare the different answers of the students and try to form a common answer for the whole class.

REFERENCES

1. D. T. Miller and D. A. Effron, "Psychological license: When it is needed and how it functions," in *Advances in Experimental Social Psychology*, M. P. Zanna and J. M. Olson, Eds., San Diego, CA: Academic Press, 2010, vol. 43, ch. 3, pp. 115–155, doi: 10.1016/S0065-2601(10)43003-8
2. D. A. Effron, "Beyond 'being good frees us to be bad:' Moral self-licensing and the fabrication of moral credentials," in *Cheating, Corruption, and Concealment: Roots of Unethical Behavior*, J.-W. van Prooijen and P. A. M. van Lange, Eds., Cambridge, UK: Cambridge University Press, 2016, ch. 3, pp. 33–54, doi: 10.1017/CBO9781316225608
3. A. Giubilini (2023), "Conscience," in *The Stanford Encyclopedia of Philosophy*, E. N. Zalta and U. Nodelman, Eds. [Online]. Available: https://plato.stanford.edu/archives/fall2023/entries/conscience/
4. H. G. Frankfurt, *On Bullshit*. Princeton, NJ: Princeton University Press, 2005.
5. J. S. Mill, *Utilitarianism*. First published in 1861. G. Sher, Ed., Indianapolis, IN: Hackett Publishing Company, 2001.
6. S. Bok, *Lying: Moral Choice in Public and Private Life*. New York, NY: Pantheon Books, 1978.
7. J. Waldron (Ed.), *Theories of Rights*. New York, NY: Oxford University Press, 1984.

6 Employment-Related Micro-Level Cases

Learning Objectives

After studying Chapters 5–7, you will have the basic skills to

- Explain the decision-making process behind one's unethical behavior using moral self-licensing
- Analyze problematic behavior and actions in small, everyday cases using moral theories
- Evaluate the consequences of identified unethical actions for both individuals and organizations

Employment-related cases differ from product-related cases because they may not be directly observable from the outside—in the same way—as missed delivery deadlines or inadequate product features. Therefore, the employment category is naturally a more intimate and less discussed class of unethical behavior. Fortunately, we were able to collect a representative set of small cases dealing with such timeless issues as discriminatory actions, termination of employment, and conduct on business trips. These should form a fertile basis for class discussions.

DISCRIMINATORY PRACTICES OF EARLY TERMINATION

There is often a probation period of a few months at the start of a new employment contract. During this time, the employer has the opportunity to examine the employee's attitudes, skills, and general suitability for the position. In addition, the employment contract can be terminated immediately without giving a reason, if necessary. The following two cases show how the flexible probation period can also be used as a convenient "back gate" for discriminatory purposes, which was certainly not one of the objectives when such a policy was introduced. But every early termination is regrettable, as finding a qualified and suitable replacement can be time-consuming. In such cases, there is no winner.

Sexual Orientation

BOX 6.1 VIGNETTE

ACTORS

Walter: boss, technical director
Willi: newly hired mechanical engineer

DOI: 10.1201/9781003485520-6

CASE

Willi, a middle-aged mechanical engineer, was hired by a hi-tech consulting company. His job description included designing ultralight structures with 3D-CAD modeling and massive finite-element simulations, as well as tutoring of other engineers to use such techniques. As was standard practice, Willi had a six-month probation period at the beginning of his employment.

During his first few months of employment, Willi established himself as a skilled and productive designer in his new team. He was also a well-liked employee among his colleagues and a good team player. Everything was just fine. Until one day, Willi's boss, Walter, heard from a reliable external source that Willi was homosexual, married to another man, and that they even had an adopted child.

This was a huge shock to Walter, who was a very conservative person. He was mad at himself for hiring Willi. Soon, Walter remembered that Willi was still on probation, and during the probation period it is possible to terminate the employment contract without giving any reasons for that—great! Walter decided to fire Willi immediately {A}.

He asked his private secretary to bring Willi into his office and told Willi coldly that his employment contract would be terminated that same day. Willi was more than surprised but took the unexpected news calmly. After this, Willi said goodbye to his fellow engineers and other employees, and left the company in a pensive mood.

First, all the employees were stunned that Willi had left their company—they had no idea for the reason of his abrupt departure. However, Walter purposely leaked the reason for his unacceptable decision to one of his subordinates, and soon everyone knew it. Walter explained he was obliged to do it because his consulting firm would undoubtedly lose clients if it became known that they do employ LGBTQ+ people {B}. After that incident, all the employees were pretty cautious about Walter. And now he had to find quickly another skilled engineer to replace Willi, which was not an easy task.

AQVM

Walter: fairness 🙁, respect 🙁, responsibility 🙁, integrity 🙁
Willi: tact 🙂

Licenses

{A} Walter's moral self-license had two optional motives: (1a) he did it sincerely for the good of his company's reputation—thus for all of them; or (1b) perhaps he had occasional homosexual thoughts in his mind and was ashamed or afraid of them. In any case, this seems to be purely Walter's personal conflict and has nothing to do with the organization's culture.

{B} Walter's moral self-license: he wanted to announce everyone (but indirectly) that he only did it for the benefit of their company, that is, for them all. Maybe he owed an explanation.

Analysis

The Utilitarian Perspective

The failure of Utilitarianism can be seen most clearly in circumstances like these. When actions are judged solely on the maximization of happiness of all affected persons, the door is opened for all sorts of (otherwise) questionable things. Now, in the case at hand, it appears that Walter's conservatism (in this instance, a euphemism for small-minded bigotry) did not result in the best overall balance of happiness over unhappiness. By firing Willi, he caused widespread uncertainty and dissatisfaction. But even if his action *had* maximized happiness among his coworkers or his clients—thereby earning a utilitarian's praise—it would still be worthy of criticism.

The main reason for Utilitarianism's failure (in this case—at least—if not in general) is that it offers an incomplete account of value. Recall, according to the utilitarian, happiness is the only *intrinsically* valuable thing there is. You and I are only valuable insofar as we are capable of experiencing happiness. Our value is secondary to (and dependent on) our capacity for feeling pleasure. But while happiness surely *is* valuable for its own sake, it is far from the *only* thing valuable for its own sake. Intrinsic value can be found elsewhere, in things like love, achievement, character, and life itself.

So, while a utilitarian would probably condemn Walter's act of firing Willi, we should hesitate to consider that the last word on the matter. For, if enough of Walter's coworkers were similarly prejudiced, the verdict would be different. But moral rightness cannot be determined by a majority, and neither can it be limited to a narrow view of value.

The Deontological Perspective

What makes a person valuable? According to a deontologist like Kant, we are valuable because we are autonomous and rational. Now, there are reasons for doubting this account of intrinsic value. For one thing, it is far too exclusive, even if we restrict our attention only to human beings. But for all its problems, Kant's view of intrinsic value does begin with a correct assumption: to understand what makes us valuable we need to look deep below the surface, past superficial differences.

The issue here is about who (or what) deserves our full moral consideration. Apparently, Walter believes that one's sexual orientation is an important factor to consider when determining who is deserving of honesty, respect, and employment. He must believe that one's self-identity (only one dimension of which involves whom one is physically and emotionally attracted to) is what makes one worthy—or not—of moral consideration. He is wrong about that.

Autonomy and rationality might not be the determining factors for membership in the moral community, but sexual orientation *certainly* is not. Walter might just as well have fired Willi for having brown eyes or disliking broccoli. Whatever it is that gives a person their value, it is not any of that. From a deontological standpoint, all that matters regarding Willi's moral standing is that he is motivated by a sense of duty and performs actions that stand up to the Categorical Imperative. By firing someone who by all *relevant* accounts is a good engineer, Walter fails in both regards.

It should be noted that open discussion of personnel decisions is prohibited by law in the United States, so firing anyone for any reason seems to be a touchy issue from the deontological perspective. But treating anyone as not deserving respect is a fundamental problem.

The Virtue Theory Perspective

This case reveals the weakness of Walter's character. When he uses the probation period to punish Willi for being gay, Walter demonstrates a lack of integrity. Instead of treating the probation period as it was intended (i.e., to see if new employees are a good fit for their position), Walter hides behind it by way of retaliating against something he personally disapproves of, unrelated to the job though it may be. Walter then fabricates a flimsy reason to justify his action. And by leaking the reason for the firing rather than being forthright about it, Walter exhibits a lack of courage, honesty, respect, and responsibility. An engineer who relies on loopholes and pretexts is neither a good engineer nor a good person.

Discussion

In many ways, this is a confusing case. Walter announced indirectly to everyone that he had fired Willi for the good of the company. But it is hard to see any *benefit* to his company from this vignette. Willi had got "up to speed" quickly, and was already a productive designer and an employee liked by his colleagues, and even a good team player. Such employees are not intended to be fired during the probationary period. So, Walter definitely caused harm to his company by firing a good mechanical engineer.

In addition, it is likely that Willi understood the real reason for his dismissal, since there were apparently no visible professional or other objective reasons for such an act. This probably led Willi to think/believe that this *company* is discriminating against people based on their sexual orientation. Therefore, the company was probably the "bad actor" for him—and Walter just did what he was supposed to do. Such a word—albeit a clear misconception—spreads swiftly in cyberspace. This could be the second harm caused by Walter.

And Walter also caused trouble for his subordinates; if a good engineer can be fired because of his sexual orientation, what other arbitrary reasons could be used in the future? With such an irrational action, he sowed the seed of *fear* among his subordinates. It apparently affected their work performance. This is the third harm caused by Walter.

As stated above in the Virtue Theory analysis, "Walter exhibits a lack of courage, honesty, respect, and responsibility." Also, he seems to be an insecure person and not a competent leader. His personality and immaturity prevent him from truly understanding what is best for his company—and eventually for himself, too. It is very likely that Walter's career path had already reached its culmination point. He was just a wicked old man. And it appears that Walter's behavior in this case had nothing to do with the ethical culture of the company. Maybe he had a homophobia [1].

Multiple Sclerosis

BOX 6.2 VIGNETTE

ACTORS

Egon: factory superintendent
Elke: project manager
Kurt: newly hired mechanical engineer

CASE

Elke, a project manager, hired a mechanical engineer, Kurt, for her project group. Kurt was an experienced engineer with a master's degree, and he had excellent recommendations. Elke was happy that she managed to find such an expert in tribology.

One evening, Elke received a surprising call from the factory superintendent Egon. She had never spoken to the "big shot" before. Egon was clearly angry: "What did you have in mind when you recruited that handicapped Kurt into your group?" Elke did not understand what Egon meant and asked for clarification. Egon said that after Kurt's routine medical examination, the company doctor had urgently informed him that Kurt had a degenerative immune-mediated disease affecting the central nervous system, better known as multiple sclerosis (MS). And thus, he could become expensive to the company if and when his disabling disease worsened {C}.

Elke did not know what to say. She thought Kurt looked as healthy as anyone, he even played frisbee golf on the recently founded company team. Egon declared that Elke must terminate Kurt's employment contract while he still has his probation period. Elke tried to resist, but for nothing.

So, after a couple of days, Elke talked to Kurt and informed him that his contract had been terminated, without giving him any reason {D}. Kurt was taken aback, but took the termination notice with tact. Furthermore, Elke found it odd that the company doctor had shared Kurt's private medical information with Egon—perhaps such was even illegal. And again, Elke's project group was missing a tribology expert.

AQVM

Egon: tact ☹, fairness ☹, responsibility ☹
Elke: respect ☺
Kurt: tact ☺

Licenses

{C} Egon's moral self-license had two overlapping parts: (1) he just wanted to save the company money in the worst possible scenario—that was his duty as the factory superintendent; therefore, (2) exclusively for the benefit of the company. Egon seemed to focus only on the possible financial implications without considering any ethical or other implications at all. He was a cold professional.

{D} Elke's moral self-license was simple: she did what she was ordered to do.

Analysis

The Utilitarian Perspective

An obvious place to focus in this case is on Egon's demand that Kurt be fired because of his medical condition. According to the account of his moral self-licensing ({C} above), Egon simply wanted to do what was financially best for his company. And by getting rid of an employee with MS, Egon was avoiding the high healthcare costs that his company could incur if Kurt's condition declined. Of course, it is impossible to know whether Kurt's health would have gotten worse, and so we can only judge Egon's action based on the consequences we know about. Kurt, to his credit, received news of his termination with grace and tact. Perhaps this indicates that he experienced relatively little unhappiness. Elke seemed uncomfortable with having to fire Kurt, but she did so anyway. All things considered, a utilitarian appears to have little to complain about regarding Egon's demand that Kurt be fired. Intuitively, that seems problematic.

Additionally, we ought to mention here the fact that Elke eased her conscience by remembering that she was simply following orders. This phrase ought to conjure up images of the horrific moral evils committed in the twentieth century by the likes of Adolf Eichmann, who took no responsibility for the Holocaust crimes he committed because he was ordered to commit them. Such a defense did not stand up to legal scrutiny because it cannot stand up to moral scrutiny (which is more fundamental). Regardless of the severity of the wrong, one cannot dodge responsibility for one's actions simply because one was "following orders."

The Deontological Perspective

Egon seems to have a very narrow and selective understanding of moral value. Kurt is not any less deserving of respect and moral consideration based on his health. As Kant's Principle of Humanity (PH) instructs, we are to treat persons with the respect they are due based on their rationality and autonomy. Kurt does not appear to be lacking in either of these areas, and so ought not to be treated by Egon as a means to an end. People are not reducible to numbers on a payroll spreadsheet.

Now, what if Kurt's rationality or autonomy *were* compromised? Would he be any less deserving of respect? Some strict interpreters of Kant's deontology might say so and would have to choose between (a) admitting that Kurt would be less worthy of respect than he had been when he was fully in control of his faculties; or (b) performing some theoretical contortionism by way of showing how Kant's theory calls for inclusion in the moral community of irrational or non-autonomous beings. Of course, option (c) could involve abandoning the theory altogether and searching for a better one.

The Virtue Theory Perspective

Egon displays a failure of character in this case. His courage, justice and compassion seem paltry, and his prioritization of material rewards over Kurt's wellbeing are regrettable. To be so devoted to money and so dismissive of human value are signs that Egon is not on the path to excellence. Anyone who attempts to cover up irrational and antisocial behavior by calling themself a "professional" has failed to understand that there is not a different set of rules for when you are on the clock from when you are off.

Elke, for her part, displayed a lack of courage when she caved in to Egon's order. Should she have stood up to Egon? Perhaps. Would that have been risky? Of course. When you defy your supervisor's orders, you may run the risk of losing your job. Maybe Elke's job is what keeps her family housed and fed and otherwise cared for. In that case, perhaps losing it would be a greater evil than compromising her values. What is clear is that each of us is responsible for our own moral character, regardless of the challenges and obstacles others present us with.

Discussion

Egon was the factory superintendent of a relatively large manufacturing plant. This plant belonged to an international corporation that engineered and manufactured industrial machinery. Egon had spent his entire career with the same factory, and the last 23 years as its director. In his current role, he had transformed an old-fashioned and inefficient factory into the company's most modern and efficient facility. It was his productive "child."

Egon was a true company man, whose leadership style was a strange hybrid of macro and micro management [2]. He constantly strived to improve his factory's manufacturing processes, but on the other hand, he also managed problematic employment issues, such as Kurt's case. Money and profit were all that mattered. And it is clear that this case was purely financial, not moral—from Egon's perspective. Hence, although we granted him the moral self-license above, he "could" have lived with a plain psychological self-license.

However, Egon's "dynasty" may have had a legal problem, because the company's doctor gave Egon *detailed* information from Kurt's private medical records. This is morally unacceptable and possibly against the privacy law. But the doctor knew the rules of Egon's "game" and followed them.

What about Elke, should she have stood up to Egon? Well, in principle yes, but in practice no. Elke knew very well that if she stood up to Egon, there could have been two fired engineers, Kurt and Elke. So, she did as she was ordered. But she acted against her values and may have got an ethical hangover.

In the analysis section, we stated that "Egon is not on the path to excellence." On the other hand, his factory seems to be on that path, at least from a productivity standpoint. Moreover, Egon would probably be willing to launch an ethics program at his factory if he was convinced that it would bring financial benefits.

QUESTIONABLE LAYOFF PROCEDURES

Layoffs are usually cumbersome in organizations, regardless of the underlying reasons [3]. They easily evoke emotions, and can be interpreted as opportunities or threats depending on the perspective: "employer versus employee." In the following, we present two layoff cases: one involving the termination of a dozen employees' contracts and another involving the firing of a single employee. These authentic cases are very different, but they both involve questionable behavior to analyze and discuss.

Clean the Corners

BOX 6.3 VIGNETTE

ACTORS

Erich: vice president of engineering
Gunther: fired computer engineer
Walter: department manager

CASE

There was a moderate recession in a European country, the gross domestic product was not growing, and many companies were laying off tens or even hundreds of employees. The overall atmosphere was rather tense and expectant. However, Erich's company was still doing pretty well due to its strategic operation on global markets. Erich was the vice president responsible for engineering; he managed a large product development center for factory automation components. He was a competent leader, whom the company's management and his subordinates trusted.

One day, while reading the latest Google News about massive layoffs at another company, Erich got an idea: people are already used to bad news, so maybe he could also "clean the corners of his development center" and lay off some of the underperforming or otherwise problematic employees. The official justification would, of course, be "production and financial reasons." He smiled to himself {E}. (Management frequently welcomes a crisis because it allows them to impose measures that would otherwise meet with resistance.)

In the next steering group meeting, he told his four department managers that they should prepare lists of problematic employees to be laid off. No more than 12 would be an appropriate number—so, an average three from each department.

Walter, one of the department managers, was especially delighted, and promptly produced three names for his list. He had been waiting for such a "cleaning opportunity" for a couple of years. One of the engineers on Walter's list had a moderate drug abuse problem, while the two others were lousy team players and poor communicators. Gunther's name was the last one on that list; he also had a somewhat difficult personality.

After a few days, Erich and Walter explained to the three selected engineers the "difficult" situation of the company and said that they were, unfortunately, being laid off for production and financial reasons {F}. Their employment would end in a month. Two of the engineers took the notice of termination meekly, but Gunther grinned strangely, for some reason.

On his last day at work, Gunther emptied his office and moved all the papers and other things downstairs to the common storage room. While he was carrying his desktop computer, it "accidentally" fell down all the stairs {G}. The computer was badly damaged and even its hard disk could not be recovered. All of Gunther's files were gone—nothing was left. In addition, Gunther had not taken regular backups of his files, even though it was company policy.

Walter understood that Gunther's computer falling was no accident, but still he did not say anything; he just wanted to get rid of Gunther as soon as possible. After this uncomfortable incident, Gunther switched to another profession outside of engineering.

AQVM

Erich: truthfulness, respect

Gunther: self-discipline, integrity, responsibility

Walter: truthfulness, respect, responsibility

Licenses

{E} Erich's moral self-license had three complementary components: (1) he wanted to increase the overall effectiveness of his development center; (2) he intended to terminate only the practically useless employees; and (3) it was solely for the benefit of the company. Erich was a pragmatic professional.

{F} Erich and Walter's moral self-license: for the benefit of the company.

{G} Gunther's moral self-license was strikingly unusual: they did me wrong, and therefore I have the moral right to retaliate—thus we are even. Gunther had, indeed, a difficult personality.

Analysis

When Socrates was presented with an opportunity to escape from prison, he declined. He had been found guilty on multiple counts and sentenced to death on flimsy, baseless grounds in a sham trial, and he knew that his imprisonment was unjust. But when his friends arranged for his escape and relocation, Socrates chose to remain in prison. Escaping would have involved breaking the law and turning his back on the rules and values that he had embraced for his entire life. As hard as it may be for the contemporary mind to understand his reasons, Socrates reminds us that two wrongs do not make a right. Perhaps that lesson is relevant to the present case.

The Utilitarian Perspective

Among the issues in this case, a few stand out. One is Erich's decision to use the recession as an excuse to "clean corners." Another is Walter's delight in being presented with an opportunity to fire problematic engineers. From the perspective of a utilitarian, it is entirely plausible to think that Erich and Walter acted in ways that maximized happiness and minimized unhappiness—particularly regarding the engineers who were poor collaborators and communicators.

Another issue that stands out is Gunther's destruction of the computer. As the case clearly implies, dropping the computer was not an accident, it was intended. An act of retribution such as this is an interesting issue for a utilitarian. On one hand, it seems proper that the act be judged simply on the basis of its consequences. Presumably, it made Gunther happy to exact some degree of revenge. And though it probably caused Walter some displeasure, he was not bothered enough to retaliate. In the end, it is not obvious that the act of intentionally destroying the computer and its files was wrong.

However, things may look different if we recall what an act of retaliation involves. Gunther *knowingly* acted in a way that would cause displeasure to those affected by his actions (Walter and Erich, in particular). This is clearly the opposite of what the Principle of Utility prescribes. According to this principle, our actions are to be judged based on the happiness (or pleasure) they bring about. But if somebody like Gunther chooses to act in a certain way because they want to bring about displeasure, could a utilitarian be anything but critical?

The Deontological Perspective

Firing employees for underperformance, illegal or unsafe behavior, and causing problems is not morally problematic, all things being equal. Our moral assessment of the actors in this case should not be taken to suggest otherwise. But deontological concerns are raised in by the way in which Erich and Walter went about firing certain employees. Both Erich and Walter wanted to benefit their company by reducing payroll and eliminating problematic employees, and they used the recession as a convenient excuse for achieving their goal. Had they confronted Gunther and the other employees directly as problems arose, perhaps it would be easier to believe that these employees were being treated not merely as means to an end (the end of profitability), but as ends in themselves. Kant's deontological theory calls on us to treat persons with respect, and it is debatable whether Gunther and the other two engineers who were fired received this sort of treatment.

One more point is worth making. For a deontologist like Kant, justice is a matter of what one deserves. Ultimately, our inherent value as persons explains why each of us deserves moral consideration. It may be worthwhile to ask yourself (and discuss with others) whether Erich and Walter deserved to be treated as they were by Gunther. Did Gunther show his employers the sort of respect they were due, or does he have a warped sense of justice?

The Virtue Theory Perspective

As we have seen in the analyses of past cases, sometimes people hide their decisions behind policies or technicalities in the hope of obscuring any moral questionability. Here, we see Erich appealing to the state of the economy in order to justify layoffs. We see Walter relying on Erich's orders by way of avoiding potentially uncomfortable conversations with troublesome employees. And we see Gunther lying about his intentions in order to get away with an act of retaliation.

In Erich's case, a failure of honesty seems evident. In Walter's, perhaps a failure of courage is to blame. And in Gunther's case, honesty is lacking, as are general levels of integrity and maturity. This is a case study involving behavior that is deceptive, petty, and vengeful. From a virtue theoretical standpoint, there is very little excellence of character to be found.

Discussion

In this particular European country, labor laws and associated regulations are quite strict, and procedures for laying off low-performing employees are fairly complex. Therefore, Erich and Walter saw the recession as an opportunity because in such conditions, it was easier to fire (problematic) employees simply for "production and financial" reasons. So, they abused this option as their company was *not* in any financial crisis. This behavior was somewhat similar to the misuse of probation period in the previous two cases.

Such happens all the time. But Erich's "clean the corners" and Walter's "cleaning opportunity" attitudes show that they were not mature professionals, as for some reason they had to relate low-performing engineers to trash. This disrespect was evident in the steering group meeting; hopefully not as a contagious model to others. There are certainly professional ways to communicate about sensitive layoff issues. And leaders at the vice-presidential level should be aware of that.

On the other hand, Gunther's irrational revenge and dropping the computer showed actually the reason *why* he was fired—he was, indeed, a difficult person. Walter ignored Gunther's revenge; perhaps it was the optimal decision under those circumstances. Besides, Gunther had not followed the rules and taken regular backups of his files. It is surprising that their IT support person had not paid attention to this omission. Hence, there was an obvious *quality issue* with the backup process of the product development center; corrective measures would be needed.

The company would benefit from an ethics program and a carefully thought-out code of conduct. In addition, Walter could shape up and become more attentive to his subordinates.

Eventually, Gunther switched to another profession outside of engineering. It may be that his unsatisfactory work performance and negative feelings were due to dissatisfaction with the engineering profession and being fired removed an internal inertia that prevented him from moving on [4].

Rude Language

BOX 6.4 VIGNETTE

ACTORS

Egon: team leader
Kurt: fired software engineer
Willi: vice president of technology

CASE

Egon was the leader of a small team that developed software for an embedded application. They used object-oriented programming (OOP) and the C++ programming language, as well as the universal modeling language (UML). It was challenging to hire skilled software designers and programmers for Egon's team because the company was located in a distant rural town.

Eight months earlier, they hired Kurt, a relatively experienced C programmer, but he had no prior experience with OOP and the UML. However, during the in-depth job interview, Kurt convinced Egon and Willi, the vice president of technology, that he was motivated and enthusiastic to learn the OOP approach and related software tools.

Months passed and Kurt did not make much progress. He was able to create just simple UML diagrams and fairly trivial pieces of object-oriented code. Furthermore, his progress had remained almost stagnant for nearly three months. Egon had discussed this uncomfortable situation thoroughly with Kurt a few times, and each time Kurt promised to shape up and soon become a productive software developer in his team. But nothing changed, he continued to underperform. On the other hand, Egon was not a particularly competent team leader in Kurt's opinion, because he had not organized proper orientation training or even tutoring for Kurt in OOP.

One morning, Willi invited Egon and Kurt to his office—he was finally fed up with the frustrating situation. After a short warm-up conversation, Willi said gruffly to Kurt: "Boy, you've been sitting on the potty for too long, it's time to let the shit out or pull your pants back up" {H}. Egon and Kurt looked at each other; they were stunned by the rude language Willi had used.

Kurt timidly replied that unfortunately he had a family issue that has been taking his attention away from work for the past couple of months, and… Willi interrupted him: "Everybody has family problems from time to time, but they don't affect work motivation. You leave this company immediately, we don't pay salaries for nothing. We will mail your personal belongings to you—you just give me your key and walk straight out!" After a few silent seconds, Kurt stood up, shook hands with Egon, gave his key to Willi and nodded, and then left feeling down.

Willi and Egon exchanged a few more words; this difficult incident was over. Egon thought Willi had a fairly big ego.

AQVM

Egon: respect 🙂
Kurt: truthfulness 😐, respect 😐, tact 🙂
Willi: tact ☹, respect ☹, responsibility ☹

Licenses

{H} Willi's moral self-license consisted of three items: (1) he knew he was a trusted executive; (2) Kurt was just a miserable underperformer who had not made adequate progress; and (3) he was acting solely for the benefit of the company. However, it is unclear why Willi chose such rude language, which was by no means common for him.

Analysis

The Utilitarian Perspective

Using rude language in professional contexts is generally a bad idea, though its acceptability probably does depend somewhat on the profession. In a professional engineering context, as this case illustrates, the use of rude or foul language can surprise, shock, and offend. A utilitarian would likely view this as the production of displeasure and would thus condemn it. Given the many other ways Willi might have expressed his dissatisfaction, the use of a rude expression was wrong in this instance.

The Deontological Perspective

In light of both the Principle of Universalizability (PU) and the PH, using rude language is something one ought to avoid doing. The PU instructs to act in ways that would allow us to accomplish our goals in a world where everybody *also* acted that way to accomplish similar goals. Assuming that one's goal when swearing is to shock or indicate disgust, we must ask if such goals would be reachable in a world where everybody used rude words. And in a world where rude words were so commonplace, it seems that their ability to convey disgust or inspire shock would be stripped from them.

The PH tells us to treat ourselves and others with respect, and it seems reasonable to suppose that offending others is inconsistent with respecting them. The use of rude words is one way of offending others, at least potentially, and so appears to violate the PH. It could be that the use of foul language runs no risk of offending when it occurs among a small group of people who are well acquainted with one another's dispositions and attitudes regarding nasty words. But when Willi had no idea how comfortable Egon and Kurt were with profanity, the risk of offending—and thereby disrespecting—was too great. As such, Willi's behavior was unjustified [5].

The Virtue Theory Perspective

There are no absolute rules against using rude language on this theory because it focuses primarily on defining good character rather than on determining right/wrong actions. The question to ask is whether swearing helps or hurts in one's sociability. Recall, Aristotle defines human goodness in terms of excellence respective of function. And because humans are by nature rational and social animals, one aspect of moral goodness is social excellence. Now, in certain social circles or situations, swearing might not threaten one's sociability. But in many others, it would. And if one is in the habit of using rude language, one is more likely to let a nasty word slip in the wrong company.

In most professional contexts, the virtuous thing to do is to avoid swearing. Depending on who you are interacting with, using rude words can be seen as disrespectful, juvenile, or unintelligent. Generally speaking, others take you as seriously as you give them reason to—so best to err on the side of discretion.

Discussion

Kurt was a relatively experienced C programmer in embedded real-time environments. That was a good starting point. However, the transition from procedural programming to OOP was hard for him. He had considerable difficulties due to the abstract nature and complexity of OOP (it really needs a change of mindset). Kurt had not received any orientation training or tutoring in this new style of programming, but Egon put him directly to the "trench." He was not ready for that, but obviously hesitated to tell. It is clear that Egon should have provided Kurt an OOP training course, and instead of leaving him alone with demanding programming tasks, some sort of tutoring would have been helpful. Perhaps Egon was too busy with his day-to-day activities, or his supervisor skills may have been insufficient. Also, it appears that Kurt had a family or other issue going on because his progress had been stalled for almost three months.

Egon seems like an inexperienced team leader and he also participated directly in the actual software development. And when he noticed problems with Kurt's progress, he probably tried to solve them by putting pressure on Kurt instead of offering him concrete help or even training. Thus, Egon himself would benefit of some leadership training—Willi should have noticed that.

The main point of this case is Willi's use of rude language or swearing. In the above Deontological analysis, we concluded that "Willi's behavior was unjustified." But why did he behave like that, even though it was not usual for him (he had a mature personality)? He was angry and frustrated, and we believe there were three reasons for that. First, he was disappointed that Kurt had not shown up as a competent programmer; the eight months with him had been a waste of time. And now they had to find another programmer to do the job, which would definitely not be an easy task. Additionally, he may have been dissatisfied with Erich in his role as team leader. Erich should have found a solution to the standard problem of converting a C programmer to a C++ programmer—it should not be that hard. Nonetheless, the use of rude language was unprofessional as well as unethical, and it did not add any positive value to this layoff issue. In conclusion, the problems with this company appear to be more managerial than ethical in nature.

IRRESPONSIBLE BEHAVIOR ON BUSINESS TRIPS

Business trips are such events during which various temptations to unethical behavior can arise. In these cases, the existing corporate culture plays a guiding role either as an "accelerator" or as a "brake." Of course, the individual always has the free will to license himself for some unethical activity—or to refuse it. Next, we discuss two fairly typical cases that demonstrate irresponsible behavior on business trips. The first one is relatively minor, while the second one is already a more serious deceptive action.

Free Lunches

BOX 6.5 VIGNETTE

ACTORS

Erich: mechanical engineer
Gunther: systems engineer
Heidi: electrical engineer

CASE

A group of young engineers—or Graduates of the Last Decade—Erich, Gunther, and Heidi, were on a three-day business trip in Southeast Australia. They worked for a major robotics and automation company in Central Europe. Their company had an ongoing joint-venture project with an Australian hi-tech firm, and together, they were developing an advanced assembly-robot cell for automotive factories.

During this work visit, some tests for a prototype cell were performed with a team of Australian colleagues in their laboratory. Erich was a specialist in mechanical structures, Gunther in real-time software, and Heidi in control electronics.

Their company had a policy that it paid a fixed daily allowance to cover meals and small everyday expenses during a business trip. That lump sum covered well all typical costs. In addition, Erich, Gunther, and Heidi worked in different departments, and their bosses had told them independently that they can offer one meal for the whole group if their hosts are also attending. The company would reimburse such minor agency expenses.

During the tiring flight from Singapore to Melbourne, Gunther got an idea: I will offer a lunch for our hosts on Tuesday, Erich on Wednesday, and Heidi on Thursday, and the hosts will most likely take us to dinner everyday. In this way, we do not need to spend our daily allowances for meals, at all. Erich and Heidi smiled and said that Gunther's idea was good, and thus, it became their mutual plan {I}.

After the busy and successful trip, they got all the lunch expenses reimbursed, as promised. And they naturally got their full daily allowances, too. Hence, the monetary benefit of this questionable activity was a maximum of $60 for each of them—this was what the three (free) lunches cost.

AQVM

Erich: self-discipline ☹, integrity ☹
Gunther: self-discipline ☹, integrity ☹
Heidi: self-discipline ☹, integrity ☹

Licenses

{I} Gunther, Erich, and Heidi's moral self-license had three parts: (1) none of them had ever before cheated their company; (2) those few lunches were cheap compared to other travel expenses from Europe to Australia; and (3) that would be a small group-serving act. How many questionable acts like this happen every day, around the world?

Analysis

The Utilitarian Perspective

It is hard to condemn the engineers' behavior in this case from a utilitarian perspective. It is not clear that their plan violated any policies, and it probably did a fine job of maximizing the happiness of everyone involved. Simple.

But it would not be difficult at all to alter this case in ways that would still pass utilitarian scrutiny, but that would pretty clearly involve wrongdoing (intuitively, at least). Without going into salacious detail, some work visits involve pleasure experienced by some at the cost of the displeasure and degradation of others. A utilitarian might weigh the happiness experienced in a situation like that against the unhappiness and determine that all is well. The ends, after all, justify the means.

There is probably a line somewhere between the innocent and the ignoble. The problem with Utilitarianism may be that it cannot draw such a line.

The Deontological Perspective

If a deontologist were to take issue with the actions of Gunther, Erich, and Heidi, it would probably involve their attempt to abuse their company's meal allowance policy. It may be that the engineers' lunch plans did no harm to anybody... but a deontologist like Kant might still object.

Consider Kant's belief regarding the moral status of non-human animals. Because they (he believed) are neither rational nor autonomous, they are not worthy of our direct moral consideration. But that does not mean we may kick dogs with moral impunity! A person who routinely kicks dogs is not doing anything wrong to the dog, *per se*. But he is cultivating a callous attitude toward pain, and may inadvertently kick another human, which would be wrong. Thus, by kicking dogs, a person does wrong *vis a vis* other persons.

What has this got to do with free lunches? Well, imagine what Gunther, Erich, and Heidi might do on their next work trip. Instead of stretching the company meal policy, they may attempt to stretch another policy... and another... and so on. By developing a dismissive attitude toward company policies while away, the engineers make it more likely that their future actions will violate a deontological principle. A moral person would strive to prevent that from happening.

The Virtue Theory Perspective

Much like Kant's caution against cultivating a callous attitude toward pain, Aristotle cautions against developing habits that prevent a person from becoming virtuous. The three engineers in this case might not have acted too extremely in their lunch-related scheming. But they may well have gone some way toward developing mischievous, deceptive, manipulative characters. And with these character traits in place, who knows what extreme actions could follow?

Again, the difficulty here is in identifying the line between acceptable and unacceptable "interpretations" of a company policy. Because of this difficulty, the engineering profession (and others, to be sure) often involves lunches, gifts, trips, and other perks.

Discussion

A common adage states that "there ain't no such thing as a free lunch." But what about the lunches of Erich, Gunther, and Heidi—were they *really* free? We will get back to this question after the following discussion.

Erich, Gunther, and Heidi were young but relatively accomplished engineers; especially Gunther who had already completed two major development projects leading to highly successful products. Gunther was a spontaneous extrovert, while Erich and Heidi were more introverted. They were on an important business trip and understood their value to the strategic joint-venture project. During the long flight there was plenty of time to think about anything. And then Gunther got a questionable idea.

He suggested to his colleagues how they could enjoy *free lunches* during their visit to Australia. The plan was simple, and such a thing could happen by coincidence, as well. Therefore, he thought that they could very well carry out the plan. Erich and Heidi agreed with him, because none of them would actually be violating any company policy. However, it is unlikely that either of them would have made such a proposal; Gunther was its sole architect. They had never cheated their company in any way. In this way, each of them saved about $60, and it was basically used to buy souvenirs at the airport. It was much needed extra money, as they all had substantial mortgages to pay off; their personal budgets were tight.

In the analysis section, we wrote: "Instead of stretching the company meal policy, they may attempt to stretch another policy… and another… and so on." Hence, there is a *threat* that because the stretching of the meal policy was so simple, such behavior could become habitual and gradually extend to larger violations. And then they, Gunther most likely, would be on an unethical trail. This is the potential price for these lunches—maybe they were not free after all.

Nightlife in Hamburg

BOX 6.6 VIGNETTE

ACTORS

Elke: senior mechanical engineer
Gunther: senior mechanical engineer
Walter: boss, department manager

CASE

Gunther was a senior mechanical engineer who represented his influential company in a standardization working group. That authoritative group developed and maintained European safety standards for certain autonomous machines. They had two physical seminars and two additional video meetings every year. These seminars lasted for two days, and they were organized on a rotating basis in different participating countries.

This year, the spring seminar was held in Bad Oldesloe, about 50 km north of the German city of Hamburg. Gunther had a flight to the Hamburg airport—named after former German Chancellor Helmut Schmidt—where he picked up a rental car and drove to the small seminar town in less than an hour.

As always, the seminar was very intensive and ended with a casual farewell dinner after the second meeting day. The next day, Gunther's boss Walter was having lunch in the company cafeteria when he received a text message from Gunther: "I missed my flight due to a huge traffic jam, but booked another flight for tomorrow morning. I'm really sorry." Walter had experienced the challenging traffic conditions in Hamburg area and was not surprised. He texted a quick reply to Gunther: "Ok, have a safe trip back home."

After Gunther's return, he had a lengthy meeting with Walter. Gunther was a little nervous, but Walter paid no attention to it. He was only interested in the outcome of the working group's meeting. Gunther was Walter's favorite subordinate and that probably caused jealousy among the other employees.

So, what is the point of this vignette? Well, several years later when Walter was already retired, he met Elke, a close colleague of Gunther's, who surprisingly leaked Walter the real reason for Gunther's missed flight in Hamburg {J}: Gunther just wanted to enjoy the wild nightlife of Hamburg—and he surely did {K}! Oh, boy—Walter was at a loss for words. He had blindly trusted Gunther.

AQVM

Elke: respect 🙁
Gunther: self-discipline 🙁, honesty 🙁, respect 🙁
Walter (first): respect 🙂
Walter (later): respect 🙁

Licenses

{J} Elke's moral self-license contained three components: (1) she was still jealous of Gunther who had been Walter's favorite subordinate; (2) she herself had always done her work very professionally; and (3) she was a loyal, dedicated employee. As Sissela Bok wrote, "to protect one's colleagues is natural; the relationship of those who work together can be very close, and the bonds that join them as close as brotherhood" [6] (p. 153). From this point of view, Elke's behavior was definitely not expected.

{K} Also Gunther's moral self-license had three complementary elements: (1) that had never happened before, although there had been quite a few similar opportunities; (2) he handled all his working-group assignment with professionalism; and (3) most importantly, he would work really hard next week to completely make up for the missed work day and the extra hotel night. Even so, Gunther's foolish act was of no actual benefit to anyone.

Analysis

The Utilitarian Perspective

Gunther's antics appear to have had mixed consequences. His wild night increased his own happiness without decreasing Walter's (until years later, anyway). But it did cause Elke a good deal of resentment and jealousy. At some point, Elke discovered Walter's secret, but seemed to feel obliged to keep it to herself (again, until years later). Even so, Elke's displeasure cannot have been of such a magnitude that she could not contain it for years. And so, a utilitarian might appear to have nothing to complain about in this case.

But the great refiner of utilitarian theory, John Stuart Mill [7], would remind us that the production of pleasure through our actions is not the only thing to worry about. We must also ensure that the pleasure we generate through our actions is of a sufficiently high quality. According to Mill, pleasures derived from engaging our "higher faculties" are of a superior quality to "base" (merely physical) pleasures. So, when we aim to maximize pleasure, we ought to do so not merely in terms of quantity, but mainly in terms of quality. And it is not clear that this was Gunther's objective during his wild night in Hamburg.

The Deontological Perspective

The fact that Gunther lied is enough to warrant deontological censure. Gunther disrespected Walter by lying to him about his reasons for missing the flight, which is a violation of the PH. Gunther also violated the PU by acting in a way that, were it to be exercised by everyone in similar circumstances, would not be sufficient to cover up his mischief. From a deontological perspective, this seems a rather straightforward case of moral failure.

Additionally, it might be argued that Elke was living up to some collegial or sororal duty by keeping Gunther's secret. But whatever codes of conduct might develop among groups of coworkers, none is more fundamental than the moral code.

The Virtue Theory Perspective

In principle, nobody should be blamed for wanting to relax and have fun every now and then, assuming one has the good sense to prevent the fun from negatively impacting one's professional (and personal) responsibilities. Gunther appears to have lacked this good sense. If he had exercised a bit of prudence, he might have avoided the disruption to his professional life. That is, he might have avoided saddling his company with the costs of a new flight and an extra night's stay in a hotel. He might also have avoided straining relationships with his colleagues. Gunther's actions in this case indicate some room for character growth and a failure in practical wisdom.

We might also wonder if Elke acted virtuously in this case. It seems that she felt the need to keep Gunther's activities in Hamburg secret… at least for the duration of Walter's time as department manager. If Elke believed that she was being loyal to Gunther, then a virtue theorist would likely point out that loyalty has its limits. By protecting Gunther's irresponsibility, Elke may well have displayed some extreme version of loyalty—blind loyalty or some such vice.

Discussion

Gunther was a bright and accomplished mechanical engineer in his forties, among the best in Walter's large engineering department. Besides, he was Walter's favorite subordinate. This may have caused jealousy and bad feelings in some of his other subordinates—Elke, the "fink," was one of those. In general, this kind of *favoritism* undermines trust in leadership and can negatively affect the work atmosphere [8].

But Gunther betrayed Walter's trust by lying about missing his flight, thus bluntly setting up the stage for a wild night in Hamburg. It cost his company one lost workday, an extra airfare, an extra hotel night, plus an extra daily allowance—about $1,000, anyway. (Un)fortunately, Walter only discovered this as a retiree, several years later.

In addition, Gunther was married and had three children. Therefore, his wild night in Hamburg might have negatively affected his personal life and wellbeing. And since every human is a *unified entity*, also Gunther's mood at work may have deteriorated, causing a decrease in work performance. Our professional and personal lives are intertwined. We only have one morale.

Gunther had even told (bragged?) about his adventures in Hamburg to his colleagues, which no doubt provoked jealousy, as it was not considered fair that the boss's "favorite boy" was allowed to do such things. But why did he tell others about this sensitive incident? Well, "experiences are worth nothing if you cannot share them with others." This is especially true in today's age of social media. As stated in the above analysis: "Gunther's actions in this case indicate some room for character growth and a failure in practical wisdom." He did wrong.

Nevertheless, such unethical behavior could have been prevented by a code of conduct expressing the company's ethical operating practices, which would have been comprehensively communicated to all employees. Being the "favorite boy," Gunther would likely have internalized almost anything his boss seemed to be committed to. Leading by example is an amazingly effective method.

"I have sinned," as President Bill Clinton apologized in 1998—what about Gunther, had he sinned too?

SUMMARY

Four of the six case studies in this chapter involve some form of *abuse* or *stretch*, related to specific acts, laws and policies:

- The employee's probationary period (Boxes 6.1 and 6.2)
- Privacy laws that protect an employee's sensitive health information (Box 6.2)

- Layoffs of employees based on production and financial reasons (Box 6.3)
- Company travel policy (Box 6.5)

In these cases, we provide the following (self-evident) recommendations to the organizations and employees concerned:

- The probationary period may only be used for the purposes for which it is intended
- No privacy laws protecting employees are violated
- "Production and financial reasons" can still be used as grounds for laying off problematic employees, but with *strict consideration* (this is anyhow going to continue)
- Employees must follow the company's code of conduct without exceptions

When these recommendations have been implemented and internalized, both employee behavior and the company's employment practices have taken a significant step forward. Today, social media can make various abuses and discriminatory behavior widely and quickly visible. Therefore, engineering managers should be especially careful when dealing with sensitive personnel issues.

The case of Box 6.4 ended up in an unfortunate (unnecessary?) layoff. This could have been avoided by arranging appropriate training and tutoring for the newly hired software engineer. After an effective training period, it is likely that all parties would have been winners—rather than losers.

Favoritism (exists in Box 6.6) is prevalent in many engineering organizations. But it should be noted that it can be harmful to *both* the favorite employee and the organization. Typically, a department manager may have a favorite employee who gets prime job assignments and benefits (flexible hours, private office, company car, etc.) that make his colleagues jealous. This can lead to a deterioration of the working atmosphere and a reduction in collective work efficiency, and also for isolating or even harassing the manager's favorite employee.

Fortunately, ethics programs would drive the companies involved toward more humane values and ethical practices in matters related to employment. That could have been of (partial) help in the cases of Boxes 6.3–6.6 for those engineers and engineering managers who acted questionably. An ethical company culture would be important, because it inhibits employees from seeing common unethical behavior as *normal*. This may make cases like those in Boxes 6.1, 6.2, and 6.5 to disappear, in the long run.

In Table 5.1, the dominant reason behind psychological/moral self-licenses is "Company's benefit" (47%), and only 20% of licenses fall into the "Other" category. But, when we moved from product-related cases to employment-related cases, the number of licenses in the "Other" category increased to 46% (Table 6.1). Hence, the motives behind unethical actions are obviously more diverse when dealing with employment issues. And the actor's company is no longer the principal motivator; instead, the actor's personality traits and even emotions play a key role.

In addition, we can make another interesting observation by comparing Tables 5.2 and 6.2. In Table 6.2, the virtue *self-discipline* is seriously violated five times, but in

TABLE 6.1
Main Reasons behind the Speculated Psychological/ Moral Self-Licenses {A} to {K}

Self-license	Company's Benefit	Actor's Benefit	Other
A			Yes
B			Yes
C	Yes		
D			Yes
E	Yes		
F	Yes		
G			Yes
H	Yes		
I		Yes	
J		Yes	
K			Yes
%	**36**	**18**	**46**

Table 5.2, this virtue does not appear at all. Why so? Well, we leave it to the reader to think about the possible reasons for this (Exercise 7 below).

Finally, we propose the idea that it might be useful to include testing the *empathy quotient* [9] as a part of the aptitude tests often used to recruit engineers and engineering managers. This would provide complementary information on the top candidates when issues related to *friendly interaction* and *natural handling of employee diversity* are considered particularly important. The cases in Boxes 6.2 and 6.4 would have benefitted from more empathetic leaders. In this way, better managers and constructive team players for multicultural engineering organizations could possibly be identified, as well.

The following Chapter 7 is similar in structure to Chapters 5 and 6, but its case studies are related to various *interaction* issues within multiple engineering companies and in one university department.

TABLE 6.2
Actor's Qualitative Virtue Measures in Employment-Related Cases

Virtue	Strength Level	Strength Level	Strength Level
Fairness	0	0	2
Honesty	0	0	**1**
Integrity	0	0	**5**
Respect	2	2	7
Responsibility	0	1	4
Self-discipline	0	0	5
Tact	3	0	2
Truthfulness	0	1	**2**

CLASS EXERCISES

1. Which one of the above cases (Box 6.1–Box 6.6) was the most against your own ethical principles? Only one justified answer is allowed.

   Class discussion: Compare the different answers of the students and try to form a single common answer for the whole class.
2. Using the following moral theoretical perspectives, analyze the behavior of the *company doctor* in the case "Multiple Sclerosis" (Box 6.2). For your information, it may be against privacy laws in that particular country to give detailed information from an individual's medical records to, for instance, the "big shot."
   a. Utilitarian perspective
   b. Deontological perspective
   c. Virtue Theory perspective
3. Construct a speculative moral self-license for the *company doctor* to be allowed to inform the factory superintendent Egon details about Kurt's MS disease (Box 6.2). In fact, this was not the only time he acted in such a way.
4. At the end of the vignette, "Clean the Corners" (Box 6.3), are the AQVMs given below:

   Erich: truthfulness, respect
   Gunther: self-discipline, integrity, responsibility
   Walter: truthfulness, respect, responsibility

   Explain the authors' possible thinking behind these highlighted virtues and their strength levels. Is there something that you would disagree with?
5. Why did Erich ("clean the corners") and Walter ("cleaning opportunity") want to show in the steering group meeting such a provocative attitude that problematic engineers are similar to trash (Box 6.3)? What did they gain by doing so in this *professional* context?

   Class discussion: In general, what do people gain by talking disrespectfully about their colleagues in workplaces? What are the typical motivations behind such derogatory behavior?
6. Form an acting pair of two volunteer students. First, they prepare a script for a role-play based on the vignette "Nightlife in Hamburg" (Box 6.6). This script is about a *variation* of the original story, where Walter had just learned from an anonymous informant (Elke?) that Gunther missed his flight on purpose to get a chance to enjoy the nightlife in Hamburg. This heated interaction takes place soon after Gunther returned to his office. The role-play is played in front of the class.

   Class discussion: What could have been the long-term consequences if Gunther had really been caught?
7. Table 6.2 summarizes the Actor's Qualitative Virtue Measures in *employment*-related cases. There, the virtue of *self-discipline* is seriously violated five (5) times. On the other hand, this particular virtue was not mentioned at all in the corresponding Table 5.2 of *product*-related cases. How would you explain this notable difference? In addition, how could someone improve his/her self-discipline? You can use one of the AI chatbots to help you with the latter question.

REFERENCES

1. S. F. Morin and E. M. Garfinkle, "Male homophobia," *Journal of Social Issues*, vol. 34, no. 1, pp. 29–47, 1978, doi: 10.1111/j.1540-4560.1978.tb02539.x
2. J. Nicholls, "Leadership in organisations: Meta, macro and micro," *European Management Journal*, vol. 6, no. 1, pp. 16–25, 1988, doi: 10.1016/0263-2373(88)90005-9
3. S. J. Ramlall, S. Al-Sabaan, and S. Magbool, "Layoffs, coping, and commitment: Impact of layoffs on employees and strategies used in coping with layoffs," *Journal of Management and Strategy*, vol. 5, no. 2, pp. 25–30, 2014, doi: 10.5430/jms.v5n2p25
4. R. E. Floyd, "Not happy? Move on!" *IEEE Potentials*, vol. 36, no. 3, pp. 23–25, 2017, doi: 10.1109/MPOT.2016.2604098
5. B. de Vries, "Is swearing morally innocent?" *Ratio*, vol. 36, no. 2, pp. 159–168, 2023, doi: 10.1111/rati.12373
6. S. Bok, *Lying: Moral Choice in Public and Private Life*. New York, NY: Pantheon Books, 1978.
7. J. S. Mill, *Utilitarianism and on Liberty: Including Mill's 'Essay on Bentham' and Selections from the Writings of Jeremy Bentham and John Austin*, 2nd ed. M. Warnock, Ed., Malden, MA: Wiley-Blackwell, 2003. (Original work published 1861).
8. J. Vveinhardt and R. Bendaraviciene, "How do nepotism and favouritism affect organisational climate?" *Frontiers in Psychology*, vol. 12, Jan. 2022, doi: 10.3389/fpsyg.2021.710140
9. E. J. Lawrence, P. Shaw, D. Baker, S. Baron-Choen, and A. S. David, "Measuring empathy: Reliability and validity of the empathy quotient," *Psychological Medicine*, vol. 34, no. 5, 2004, pp. 911–920, doi: 10.1017/S0033291703001624

7 Interaction-Related Micro-Level Cases

Learning Objectives

After studying Chapters 5–7, you will have the basic skills to

- Explain the decision-making process behind one's unethical behavior using moral self-licensing
- Analyze problematic behavior and actions in small, everyday cases using moral theories
- Evaluate the consequences of identified unethical actions for both individuals and organizations

Effective and sincere interaction between individuals forms the basis of functioning organizations. When the standard rules of professional interaction do not seem to be enough, unethical behavior is sometimes seen as an alternative—either consciously or subconsciously licensed by the actor. In these unfortunate cases, virtues such as fairness, honesty, and respect are often at risk. Fortunately, this type of behavior is increasingly *not* considered an option in advanced organizations. Many companies today are becoming aware of ethical issues; mere compliance with laws is no longer enough, and the importance of ethical values also in engineering organizations is increasing.

In this final chapter of micro-level case studies, we present a representative set of interaction-related cases that involve some degree of unethical behavior. It should be noted that the end result of such behavior is often a weakened work atmosphere, which, in turn, may reduce organizational effectiveness. Additionally, interpersonal trust may be undermined.

A STRESSFUL LEGAL INTERVIEW

Even technical experts may sometimes get involved in legal interviews or lawsuits in their organizations. These can be stressful experiences for average engineers who represent their companies in bothersome disputes. It quickly becomes apparent that interacting with lawyers and judges is very different from interacting with fellow engineers or engineering managers. And distressing thoughts, like "Have I done something wrong after all?" might creep into one's mind. The following case deals with an incident in a legal interview, where a committed engineer gave a few dishonest answers solely to protect his company.

 DOI: 10.1201/9781003485520-7

I Don't Recall

BOX 7.1 VIGNETTE

ACTORS

Erich: electrical engineer
Heidi: technology attorney
Walter: boss, vice president of engineering

CASE

Erich, a young and ambitious electrical engineer, worked in the field of industry applications. He was employed by a European corporation, which transferred him to their U.S. subsidiary company. After completing his tenure in Salt Lake City (SLC), Utah, with the subsidiary, Erich was transferred further to a sales office in Seattle, Washington; there he served as a technical advisor in the company's sales team.

He had done an excellent job while in SLC and had received the Company Award of Merit with the citation: "For providing outstanding support in achieving our objectives." Erich was proud of this distinction, and his career was on the rise. But his excessive loyalty raised certain ethical concerns.

During his stay in Seattle, a brand-new product was sold to a domestic customer. However, it soon became apparent that the (aspirational) promises made and the customer's expectations did not align, and the customer was disappointed with that product. Ultimately, the client company turned the case into a legal issue due to contract violations and misuse of the phrase "well proven." In fact, the new product was launched for sales prematurely, but this was known only to a trusted core group, including Erich.

A few years later, when Erich had already returned to the parent corporation, two lawyers from the client company unexpectedly invited him for an interview. They wanted to ask questions about the "Seattle incident" and get an official statement from Erich; as he had been employed by the Seattle office at the time of the incident.

This was a bothersome issue for the corporation, and they hired a top attorney, Heidi, to counsel Erich before and during the interview. The attorney spent one intensive day with Erich prior to the interview and coached him on how to respond appropriately when interviewed. Below are some of Heidi's general advice {A}:

1. Do not answer broad questions but always ask for specifics.
2. Briefly answer the questions; do not tell anything that was not specifically asked.
3. If a true answer would put your company in a negative light, say that you do not remember (because it happened a few years ago).

4. Play a bit dumb and slow when you give your answers.
5. Only use your native language (not English)—a professional translator acts between you and the client's lawyers; this makes the interview process less convenient for the interviewers.

The rigorous interview lasted five hours and it was a stressful experience for Erich; he was exhausted. But Erich acted just as he had been trained {B}, and both Heidi and his superior, Walter, praised Erich's high level of professionalism and total loyalty {C}.

Nevertheless, Erich did not feel comfortable at all; he knew some of his answers had, indeed, been dishonest or even lies. On the other hand, Erich was 100% committed to his company and saw no alternatives. And his future dream was to become a CTO in one of the subsidiaries of that worldwide corporation—maybe in Australia.

AQVM

Erich: honesty 🙁, respect 🙁, loyalty 🙂
Heidi: truthfulness 🙁, respect 🙁, loyalty 🙂
Walter: truthfulness 😐, respect 😐, responsibility 🙁

Licenses

{A} Heidi's routine moral self-license consisted of two parts: (1) this is just normal practice in legal interviews; and (2) for the benefit of her client corporation. She had given herself a similar license numerous times before.

{B} Erich's moral self-license had three complementary elements: (1) he was advised to do so by a respected technology attorney whom he regarded as an authority; (2) he had no viable alternatives; and (3) for the benefit of the corporation. Erich apparently violated his personal ethical standards.

{C} Walter's moral self-license was simply: Erich's dishonesty was a strong sign of loyalty and was done as an altruistic service to the corporation, it benefitted us all. Walter was obviously proud of his subordinate.

Analysis

The Utilitarian Perspective

Heidi, the attorney who coached Erich, may have utilitarian considerations in mind. Her focus is on using Erich to create the greatest benefit for the company, and her methods involve whatever actions—including lying—are necessary for securing that benefit. As we know, the utilitarian is likely to regard any means as justified, so long as they bring about a desired end (involving the production of happiness or interest satisfaction). But if this is how Heidi is thinking, she is not being as diligent a utilitarian as she could be. For, she appears to be prioritizing her interests and those of the company that employed her. She is putting her and their happiness ahead of the happiness of the client company and its lawyers.

You may be thinking that is entirely appropriate, considering the fact that Heidi was hired to protect just one set of interests. Worrying about the happiness of the client company is not in Heidi's contract, so why should she give it a moment's thought? But recall, there is no justification on a utilitarian framework for putting one set of interests above others—at least in cases like the one under consideration. Happiness is happiness, no matter who experiences it. And if Heidi (through Erich) works to frustrate the interests of the client company, she might not be doing all that she could to maximize the happiness of everyone affected by her actions. Besides, appealing to a legal contract is no escape from moral responsibility.

Perhaps, too, there is an element of moral relativism at work in Heidi's mind (see Chapter 3 for a discussion of moral relativism). By licensing her actions with what is "normal practice" in legal interviews, she appears to believe that right and wrong are simply a function of what a group happens to accept or reject. So, if deception and obfuscation are commonly understood to be normal behavior in legal interviews, then behaving as Heidi advised would not be wrong. Similarly, says the moral relativist, if cheating, hazing, bullying, and discrimination were accepted by a community of professionals, those actions should not be considered wrong in that community. However, as we have more fully explained in Chapter 3, the simple fact that something is accepted or commonplace is not enough to make it morally right. Moral relativism is a deeply flawed attempt at a moral theory, and thus deserves no more of our time. Suffice it to say, if Heidi is operating under the assumption that right actions are determined by what is "normal practice," she needs to think harder.

The Deontological Perspective

Focusing on Erich, we can identify some deontological concerns. For one, he acted in ways that he knew to be dishonest. For another, he was motivated by his professional goals rather than by a pure sense of duty. A bit of reflection will show that by heeding the advice of the attorney, Erich acted in ways that violate both the Principle of Universalizability (PU) and the Principle of Humanity (PH). In short, Erich sought to achieve a goal through unjustifiable means and, in the process, disrespected others by ignoring their rationality and autonomy.

The Virtue Theory Perspective

Having looked at the actions of Heidi and Erich, let us focus now on Walter. The main issue involves his understanding of the virtue of loyalty. As we have discussed elsewhere [1], loyalty is a virtue that corresponds to two vices: disloyalty and zeal. A person who is disloyal acts in a way that reveals a deficiency, a lack of loyalty. A person who is zealous acts in a way that reveals an excess, or "too much" loyalty. Now, in order to determine whether one has been loyal or not, it is necessary to consider both the person performing the action and the "recipient" or "target" of the action. And if that target is unworthy of one's loyalty, then one's action toward it cannot be considered virtuous.

Consider a boss who (1) embezzles from his company and (2) asks an employee in accounting who becomes aware of the embezzlement to "fix" the numbers in reports to cover up the misdeed. That boss is doing something wrong, and is therefore unworthy of the positive regard of the accountant. And if the accountant decided to follow the boss's instructions, the resulting action would not be a loyal one—it would be an extreme act, an instance of being "too loyal." Similarly, by praising Erich for having lied to the client company's lawyers, Walter is mislabeling the action. Erich probably protected a sales manager of his company who made false promises to a customer, and who therefore is not worthy of Erich's protection. So, by lying to protect an unworthy entity, Erich behaved viciously, not virtuously.

One final point here: a virtue theorist like Aristotle would not insist that it is *always* vicious to deceive. To deceive in order to protect a *worthy* entity could well be justified. But to deceive in order to protect a sales manager that lies to clients does not count as loyal.

Discussion

Erich was devoted to his company and its success. And he considered the advice of the attorney, Heidi, as unquestionable instructions from an *authority*. Therefore, during the stressful legal interview, he quite altruistically chose the dishonest path. He was still pretty "blue-eyed." In addition, he was motivated by his long-term career goals rather than a mere sense of duty. But he did pay a price for his unethical behavior; he apparently suffered from an ethical hangover.

Heidi was a top attorney who knew how to play the "game." She performed her professional role as expected by her client company; dishonesty and lying were among her counseling tools. With lawyers in similar situations in mind, Sissela Bok asked a pondering question: "Can it be argued that such lies are so common by now that they can form an accepted practice that everyone knows about—much like a game or bargaining in a bazaar?" [2] (p. 163). Hopefully, this is not the case.

Erich's boss, Walter, was a loyal pragmatist himself, and he sincerely thought that Erich's dishonest behavior was loyal and right. In praising Erich for having lied to the client company's lawyers, Walter was *mislabeling* the unethical behavior. In general, this type of mislabeling should be avoided, as it can reinforce one's tendency toward dishonest actions—especially among young and inexperienced professionals. And because it is wrong.

SUPERIOR–SUBORDINATE ISSUES

Mutual trust is a necessary basis for successful superior–subordinate relationships. And the three principal factors behind such trust are *ability*, *benevolence*, and *integrity* [3]. In addition, the development of trust through positive interaction can be understood within the framework of social exchange theory [4]. But trust is a dynamic variable that can either grow or diminish, depending on the behavior of each party. In the following, we present four vignettes dealing with negative interactions between superiors and subordinates, which can even lead to a loss of mutual trust.

Executive's Early Warning

BOX 7.2 VIGNETTE

ACTORS

Erich: research and development manager
Walter: executive vice president

CASE

Erich worked in the research center of a large electronics and software company. Recently, a small satellite unit of the center was established in a technology and innovation park. This new unit intended to focus on research in the field of consumer electronics, which was one of the main divisions of the company. The digital high-definition television (HDTV) technology was going to be their primary research area.

Erich managed to recruit four top engineers to his team. The challenging HDTV design project attracted these bright individuals. They were experts in image processing and microelectronics, and wanted to work with that cutting-edge technology. Initially, they were assigned to ongoing projects for traditional digital televisions. The big HDTV project was supposed to begin in about a year. Erich's team members quickly learned the principles of digital television and became productive members of their projects. And every time Erich held a team meeting, those guys would eagerly ask about the awaited HDTV project and its schedule. Months passed, but Erich had no HDTV news to share with his hardworking team.

Every year, the consumer electronics division had a recreation day for its research and development staff. This summer, that long-awaited day was at a beautiful holiday village by a mountain lake. There were all kinds of group games, an abundant and delicious barbecue with plenty of beer and wine, and finally, a relaxing sauna; people were having a good time.

Very late in the evening, Erich went to the sauna. It was almost empty, only one man was sitting on the upper boards, apparently drowsing. Erich sat down next to the man and realized that he was Walter, the executive vice president of the division. Between them was an empty wine bottle and a knocked over beer can. Walter woke up and began to talk this and that. Suddenly, he asked why Erich was working for the division? Erich thought for a moment and answered that currently he was especially looking forward to the upcoming HDTV project.

Walter stared at Erich and slurred that there will be no HDTV project—this whole division will be sold to Asians {D}. Erich did not know what to say, he wanted to ask more about it but did not dare. So, he just threw water on the hot stones of the stove, the humidity rose rapidly, it got really hot. Walter dropped asleep. For some time, Erich thought about what Walter had said; were they just meaningless words of a drunken executive, or could it be that the consumer electronics division was really up for sale?

Over the next few weeks, Erich thought almost daily of Walter's foreseeing words. If the HDTV project does not come at all, his motivation to stay in his company would certainly diminish. Finally, he decided to contact his previous employer and ask if they had something interesting to offer him. They had, and Erich accepted their generous job offer. He left his current job and did not tell his stunned team members the real reason for his unexpected departure {E}.

Almost a year later, he heard news on his car radio that the consumer electronics division was being sold to a Chinese company. In the end, the entire product development of televisions was concentrated in Shanghai. Erich was glad he left his job soon after Walter's "early warning." But his four team members were very disappointed—they felt they had wasted a couple of years of their lives waiting for the unrealized HDTV project. They also suspected that Erich had known something about the sale of the entire division when he left.

AQVM

Erich: fairness, respect
Walter: self-discipline, responsibility

Licenses

{D} Walter's moral self-license had two components: (1) he was pretty drunk and thus had lost his usual sense of control and responsibility; and (2) he was personally bitter about those selling plans—what would his own future look like? Under normal circumstances, Walter would not have revealed such company secrets.

{E} Erich's psychological self-license: he could not be sure if Walter's "early warning" was true and he did not want to spread baseless rumors. Erich was not a teller of vague gossip.

Analysis

The Utilitarian Perspective

An obvious place to focus is on Erich's decision to leave the company. More specifically, we ought to ask if the way he acted on that decision stands up to utilitarian scrutiny. After all, we cannot blame Erich for having been tipped off by the sweaty Vice President. Walter's unprofessional conduct is no fault of Erich's. But perhaps the disappointment and frustration experienced by his team members *are* Erich's fault. Deciding to act on Walter's admission, Erich found a better job with a former employer and wound up happier than he would have been had he stayed put. Could he have done anything to bring about more good? Certainly, he could have shared the (admittedly uncertain) news with his team members, who could then have searched for better opportunities themselves, instead of feeling stunned, disappointed, and resentful toward Erich and the company. As it turns out, that would not have caused a great deal of dissatisfaction at the company, since the consumer electronics division was to be sold anyway. So, by focusing his attention only on his own interests, Erich has failed in light of the Utilitarian standard.

The Deontological Perspective

When looking at Erich's actions in this case as a deontologist, we should ask whether he performed only those actions that he was obliged to perform. Two problems stand out. First, did Erich show Walter the respect he deserved? While it is true that Erich did not solicit a drunken Walter for information, Erich did take that (presumably confidential) information and use it to his own advantage. A deontologist may worry that this amounts to treating Walter as a means to an end. Better, perhaps, for Erich to meet privately with Walter a few days later to discuss the matter and give Walter a chance to disclose (or not) the company's plans. Second, did Erich do right by his team members? By withholding potentially significant information from them, he constrained their ability to act autonomously. In other words, he prevented them from making informed decisions about their careers. Doubtless, Erich wanted to avoid spreading rumors—especially if there was a chance that they could be false. But if he had been forthcoming with his team, and then followed up with Walter, perhaps Erich would have garnered the approval of a deontologist.

The Virtue Theory Perspective

Let us start with the obvious: Walter did not behave virtuously by getting drunk and revealing company secrets in sauna. This is not to say that a virtue theorist would never approve of drinking alcohol. It is simply to point out that had Walter exhibited moderation, he would have refrained from drinking *to excess*. He would have enjoyed himself without behaving unprofessionally and recklessly. After all, drinking alcohol in sauna does fly in the face of medical wisdom. And the fact that Walter behaved this way outside of working hours does little to help the matter. When one is a professional, one represents that profession all the time. Walter may have been "off the clock," but his actions reflect poorly on himself, his company, and his profession. It is an awfully big responsibility that professionals—engineers included—bear, but that is the reality of it.

Now, what about Erich's decision to leave the company? That appears to be perfectly reasonable from a virtue theory perspective. His company's loyalty toward him seemed doubtful, and so Erich's decision to seek another job should not be considered disloyal. What might be questionable, again, is the decision to hide the news of the sale of the division from his team. This could be read as disloyalty toward the team members, or perhaps a lack of honesty or courage. In any case, Erich's actions would likely elicit a mix of praise and blame from a virtue theorist.

Discussion

Walter was a respected executive and he had always behaved professionally in front of the company's employees. This sauna incident was a unique exception; he was drunk and thus his inhibitions were lowered and his judgment impaired. But why did he allow himself to get that drunk at the division's recreation event? Probably because he was very frustrated with the aims to sell the consumer electronics division to a foreign company. This would likely mean that his comfortable executive position would be in jeopardy. And he just wanted to share with someone what was bothering him; thus, he revealed company secrets to Erich in sauna. That was unprofessional.

It was kind of a shock to Erich to hear that the consumer electronics division was for sale. If this were to happen, it would naturally mean organizational changes that could also affect his position and project assignments. But Walter's "early warning" was *not* an official announcement—Erich was *not* sure that was going to happen.

Hence, Erich decided to save himself and found quickly a new job, but because he did not want to spread any rumors, he did not tell his team members about the possible sales plans. This was an ethical dilemma.

The four members of Erich's team were the innocent victims of this incident. Erich had managed to recruit ambitious and bright engineers to his team because of the promised HDTV project. But, in the end, they were disappointed when the consumer electronics division was sold to a Chinese company and the entire product development of televisions was concentrated in China. They were fired before getting involved with the tempting challenges of the HDTV, and they were resentful of Erich who had "left the sinking ship" without any warning. What else can we say to this? Well, not much—such happens all the time.

I'm Gonna Quit

BOX 7.3 VIGNETTE

ACTORS

Egon: project manager
Erich: junior engineer
Gunther: senior engineer

CASE

Erich, a junior electrical engineer, worked on Egon's project, which developed an intelligent control system for process automation applications. After relatively extensive study, he proposed a new communications technology to be used for low-latency message transfer between intelligent nodes. Egon was unfamiliar with that technology and opposed its use. He thought that the well-proven technology used in the current product should also be used in this new system.

Erich said it does not make sense to use that oldish technology in a product that will still be in production seven years from now. Moreover, his suggested technology would offer a smooth evolution path for future product generations, as well. This debate went on for some time, and eventually turned into an argument. Finally, Egon said that he was the project manager and it was his job to make such decisions, not Erich's {F}. But Erich was determined. If you force to use that old technology, I am not going to work on this project anymore—I am going to quit {G}! Erich's trust in Egon had suddenly weakened. Egon glanced at his watch and left; it was Friday afternoon.

Gunther, a senior engineer, had listened to Egon and Erich arguing. He told Erich that there is no use to argue with Egon, just say "yes, yes"—and then do what you think is best and right. Egon no longer understands technical details, he simply lives with milestones and budgets. Gunther himself had used such a flexible practice for years, and it had worked pretty well {H}. There had been only a couple of "bumps" between Egon and Gunther. But Erich did not want to do things behind anyone's back. Thus, he was not going to adopt Gunther's unconventional practice.

On Monday morning, Egon and Erich met at the coffee machine to have their usual cappuccinos, and Egon said in a friendly way that they would adopt the new communications technology Erich had proposed. Accordingly, the earlier dispute was settled. From then on, the collaboration between Egon and Erich continued smoothly. They seemed to trust and respect each other again.

AQVM

Egon (first): fairness ☹
Egon (later): fairness ☺
Erich: tact ☹, determination ☺
Gunther: integrity ☹, respect ☹, responsibility ☹

Licenses

{F} Egon's psychological self-license was pragmatic: (1) he just wanted to minimize the possible schedule and other risks of the project; and (2) solely for the benefit of the company. Egon only wanted the best for his project.

{G} Erich's moral self-license had two parts: (1) he was objective, he knew and understood his proposition very well; and (2) for the benefit of the company. Erich surely was no negotiator.

{H} Gunther's moral self-license consisted of four elements: (1) he did not like Egon at all; (2) he was proud of his own achievements; (3) he considered himself independent; and (4) he did it just for himself. Gunther was not a constructive team player.

Analysis

This case does a fine job of highlighting the fact that being a professional engineer involves more than just the technical side of the job. Like any profession, it also involves *interpersonal* matters, which underscores the importance and unavoidability of doing ethics.

The Utilitarian Perspective

By initially refusing to adopt new technology, perhaps Egon was a bit too focused on his own interests. Learning new things can be difficult, and Egon may have had his own happiness in mind when he dismissed Erich's suggestion. For his part, Erich made a very reasonable proposal regarding the communications technology. Updating it would allow the product to be useful well into the future, which seems like a long-term benefit. A utilitarian would likely favor Erich's long-term approach and criticize Egon's short-term (and apparently self-interested) approach.

But what about the argument that ensued? What about Erich's outburst? Problematic though these may seem, they brought about a smooth resolution to the situation. Arguments are not pleasant, but the one between Egon and Erich seems to have served a purpose. It allowed Egon and Erich to voice their opinions and gave them something to ponder over the weekend. Not every argument will resolve so nicely as this one, of course. But "all's well that ends well."

The Deontological Perspective

Let us shift our attention to Gunther's part in this case. Having overheard the argument, he advises Erich to pay lip service to Egon going forward. In other words, Gunther suggests that Erich should tell Egon what he wants to hear, even if it is not what Erich is actually going to do. Why does Gunther recommend this? Because it has "worked" in the past, only resulting in a couple of disagreements. This seems like a fairly consequentialist outlook, and one wonders if Gunther values ends over means or, worse, expediency over rightness.

Also worrisome is Gunther's disregard for Egon. Gunther's attitude toward Egon is dismissive and condescending, and certainly would not be considered respectful. Showing Egon the respect he deserves as a project manager (and certainly as a person) would involve an honest discussion and, when necessary, a civil exchange of opposing viewpoints. Gunther seems all too eager to bypass this and take the path of least resistance.

The Virtue Theory Perspective

If one character trait stands out in this case, it is stubbornness. Egon exhibits this trait when he refuses to incorporate the new technology. Erich exhibits it when he refuses to utilize the old technology. Naturally, it is not wrong to have a viewpoint supported by reasons. And it is good to be willing to consider opposing viewpoints and the reasons supporting them. Both engineers appear to have some support for their views—Egon claiming that well-proven tech should be used, and Erich suggesting that future-ready tech would make sense. If they had engaged in a genuine exchange concerning the comparative values of their positions, they might have avoided an argument. But they do not appear to have listened to each other. Egon's stubbornness caused him to put his foot down and appeal to his authority. And Erich's stubbornness led him to threaten quitting. That seems rather too immoderate to garner a virtue theorist's approval.

Then again, after a few days of cooling off, Egon and Erich appeared to have put their differences aside. One wonders if their argument served as a form of *catharsis*—the word that Aristotle would have used to describe a sort of emotional cleansing, a release of negative feelings, that one can achieve through various actions and experiences.

One last point: Gunther does not appear to be a very virtuous person. His attitude toward Egon reveals a lack of honesty and magnanimity (a virtue relating to what we might call "integrity"). By contrast, Erich does well to ignore Gunther's advice. By doing so, Erich displays a virtuous character.

Discussion

Egon, an experienced project manager, and Erich, a junior engineer, shared the cumbersome character trait of stubbornness; Egon's stubbornness made him put his foot down and appeal to his authority, and Erich's stubbornness made him threaten to quit. Instead, they should have sat down calmly and discussed the communications-technology issue *objectively*. This is how professional engineers are expected to interact. Fortunately, they also had another trait in common; they were able to calm down and end up to a rational solution—and *forget* their past arguing. Nevertheless, they could both benefit from human-interaction training to hone and practice their skills and effectiveness as a leader and team member. Maybe they could even be sent on the same course.

Gunther's own practice and his unconventional advice to Erich when dealing with Egon showed a lot about Gunther's personality. As noted in the analysis above, he "does not appear to be a very virtuous person." But why did he treat Egon in such a disrespectful way and even want to steer Erich on the same pathway? It is fair to open the door a little to the past. Several years before, both Egon and Gunther were recruited for a demanding product development project. Both of them were skilled experts, and the department manager had to make a difficult decision: which one of these new engineers should be assigned as project manager? In the end, he chose Egon, and Gunther became "just" a member of Egon's team. From those early days, Gunther had been bitter toward Egon and never respected him; there was constant tension between these two professionals. This was of no benefit to anyone, and certainly not to their joint projects. Unfortunately, Gunther seems to be such a strong personality that it is hard to imagine any cure for the underlying problem. He probably thought that Egon got what he "deserved."

C Compiler Version 4.2

BOX 7.4 VIGNETTE

ACTORS

Egon: consulting engineer
Elke: vice president of engineering
Willi: laboratory manager

CASE

Egon was a consulting engineer with a Ph.D. degree from an Ivy-League university. His expertise was in digital signal processing (DSP) algorithms and their efficient implementation in signal processor environments. He had a six-month contract with an engineering firm, where he was going to develop DSP software for a sophisticated measurement system to be used in their laboratory.

Elke, the vice president of engineering, had hired him for this special project, because they did not have any expertise with adaptive sensor fusion and related algorithms {I}. However, their laboratory manager Willi would have liked to develop this DSP software himself. He had just completed a one-semester DSP online course at a local college. Thus, Willi was very disappointed when Egon was hired to do "his job." That decision clearly undermined Willi's trust in his boss—and he felt sorry for himself.

On Egon's first day in office, Elke introduced him to Willi and told that Willi will be Egon's supervisor for the duration of the contract. Willi said nothing, just nodded. After Elke left, Willi showed Egon his office and gave him the office key and the password for the computer. There was a palpable tension between Willi and Egon.

When Egon logged on to his computer, he noticed that the version of the C compiler he was supposed to use was 1.3. Egon was amazed; that primitive version was more than three years old. He was well aware that it did not use the circular addressing mode of the processor for implementing delay lines but a slower for-loop instead. Also, that ancient compiler did not utilize the parallel instructions of the CPU effectively. On the other hand, the latest existing version was already 4.2 and it generated relatively efficient object code.

Egon wanted to get the latest version and went to see Willi. However, Willi told that they did not have any newer version of the compiler {J}; thus, Egon had to work with the old version that was on his computer. Egon was frustrated. He was not sure if Willi had been telling the truth because Elke had told him that Willi was truly meticulous about the computers and instruments in his lab. For this reason, software versions should also be up-to-date.

A few days later, when Egon was working very late, the janitor came to empty his trash can. And after Egon's office, she continued to Willi's office. Suddenly, Egon remembered seeing two CD storage boxes on Willi's bookshelf; maybe those boxes contained a newer version of the compiler. Egon followed the janitor into Willi's office and said he would lock the door when he left.

He opened the first storage box and found only user's manuals as well as component and instrument catalogs, no software disks. But the second box contained several software installation CDs. And, yes, also the missed C compiler version 4.2 was there! Egon took the disk and installed the latest version on his computer—he was feeling good {K}.

The next morning, Egon went to see Willi and innocently told him that he had found the C compiler version 4.2 CD on Willi's bookshelf. Willi's face turned red, "how did you get into my locked office?" Egon said he got in when the janitor was emptying Willi's trash can last night. Willi took the CD from Egon and put it back in his box. He said nothing more. Willi thought that Egon was Elke's protege.

AQVM

Egon: integrity, respect, responsibility
Elke: responsibility
Willi: honesty, respect, responsibility

Licenses

{I} Elke's psychological self-license was understandable: (1) algorithms for adaptive sensor fusion are definitely not basic DSP; (2) Willi had enough other work to do in the lab; and (3) for the benefit of the company. Elke saw that Willi was not yet ready to work with advanced DSP algorithms after only taking a basic DSP course.

{J} Willi's moral self-license was a bit naive: he was still bitter that Egon had taken "his job," so he protested against it. Maybe Elke should have communicated this sensitive issue more carefully to Willi.

{K} Egon's moral self-license reflected his straightforward personality: (1) he was frustrated by the situation; (2) he somehow suspected that Willi had the latest version; (3) he got a convenient opportunity to check; and (4) for the benefit of the company. What else could he have done?

Analysis

The Utilitarian Perspective

It is difficult to assess the decisions and actions in this case as a whole, since we do not know the overall consequences that resulted from Egon's time with Elke and Willi's company. What we can say, generally, is that although there are some questionable actions in this case, they are redeemed in the eyes of a utilitarian if they produce the best possible balance of happiness over unhappiness.

What are these questionable actions? One is the lie that Willi told to Egon concerning the compiler version. It is difficult to imagine how this could have benefitted everybody involved. Indeed, the lie seems to have been a result of Willi's negative feelings toward Elke and Egon, told only to make himself feel better—and perhaps sabotage Egon's efforts to some degree.

Another questionable action involves Egon accessing Willi's office. While this pleased Egon, it certainly did not please Willi. And based on the description of their reactions in the case, it seems fair to say that although Egon was made moderately happy, Willi was made extremely unhappy. So, while the overall situation might have resolved itself to a utilitarian's satisfaction, there are consequences along the way—intermediate outcomes, shall we say—that would cause a utilitarian concern.

The Deontological Perspective

If we focus on the two actions examined above (Willi lying to Egon and Egon accessing Willi's office), we find that they raise concerns from a deontological perspective. After all, by telling a lie in order to exact some sort of revenge, Willi endorses a maxim that is not universalizable. He also undermines Egon's autonomy, which goes against the PH. Similarly, Egon acts on a non-universalizable maxim by accessing Willi's office without permission. And in order to do so, he treats the janitor as a means to an end. The manipulation and deceit in this case would raise a lot of red flags for the deontologist.

The Virtue Theory Perspective

Modesty is a virtue involving an accurate assessment of one's abilities, qualities, and value. Thinking too little of oneself is a vice, as is thinking too highly of oneself. Perhaps Willi is guilty of the latter when he regards the DSP software development as "his project." His experience is limited to one college course, while Egon's experience is much deeper. By overestimating his own abilities, Willi exhibits vice that leads to resentment and immoderate actions.

By the same token, perhaps Egon feels a bit too entitled in his role as DSP expert. He did the right thing by first consulting Willi about the out-of-date software. But when he reached a dead end with that conversation, Egon acted deceptively and boldly. He might instead have spoken with Elke, who might have resolved the situation rather more honorably. But as things stand in the case, neither Egon nor Willi displays a very virtuous character.

Discussion

Willi was a pedantic laboratory manager who loved his job. He had recently taken a DSP course, and was looking for opportunities to utilize this new knowledge and develop his skills in DSP technology. And he saw the adaptive sensor fusion project as an excellent opportunity for him. But his professional pride was badly hurt when Elke gave the project to a consultant, Egon. Probably Willi did not understand the complexity of that project—it was certainly not basic DSP. He protested against Elke's decision and wanted to make Egon's life somewhat difficult by *not* giving him the latest compiler version—he behaved immaturely by lying to Egon. Normally, Willi was very loyal to his company; this incident was a rare exception.

Egon was an expert in DSP and had Elke's full support for the assignment. He seems to be a task-oriented professional for whom other people were more or less objects that he used for his purposes. Egon was willing to use any means to get the job done, and he also set high standards for the quality of the results. No one would stand in his way. He did not respect Willi, even though Willi was his superior. With this background information, it is no surprise that Egon used the janitor to enter Willi's office and "borrowed" the software CD from the storage box—he probably saw nothing wrong with what he was doing. Maybe he thought a little too much of himself.

What went wrong already at the very beginning of this micro-level case? Elke should have handled the case more smoothly, showing respect for Willi and his newly acquired piece of knowledge capital. Since Willi was a loyal employee, he certainly would have understood the appointment of a consultant if Elke had explained the situation to him objectively without downplaying Willi's basic DSP skills. Additionally, Willi could have had some sort of observer role in this project, which would have been a good learning opportunity for him. And after such preparatory actions, this unethical case would hardly exist at all. Meticulous superior–subordinate interaction is, indeed, of great importance [5].

This Is Extortion!

BOX 7.5 VIGNETTE

ACTORS

Egon: director of the business unit
Erich: industrial engineer
Walter: boss, department manager

CASE

Erich, an ambitious engineer in his late twenties, trusted his competent and benevolent boss, Walter. During the past few years, Walter had given him interesting and challenging job opportunities. They had a very good relationship, nearly friendship. Sometimes they even went to basketball games together.

Just before Christmas, Walter and Erich were on a business trip from Amsterdam to Groningen by train. The train ride took about two and a half hours. They were chatting about this and that. Erich would eagerly like to have a multi-year assignment to one of the company's foreign subsidiaries; he thought that would do good for his future career and personal life. Therefore, he openly told Walter that if their company could not offer such an opportunity for him, he will probably leave to another company where he could get such a foreign assignment {L}.

Walter glared at Erich and angrily said that he was extorting—revealing a great deal about his personality. Erich was surprised of Walter's unreasonably strong reaction. He calmly continued, that actually, you did the same thing before your recent two years in Japan. Walter exclaimed that Erich's absurd claim was a lie, he would never do anything so low! How do you even dare to think of something so absurd {M}?

Erich told Walter that a couple of years ago, under special circumstances, the director of the business unit, Egon, had told him that Walter had used a kind of extortion when he wanted a temporary assignment to the company's Japanese subsidiary {N}. So, we both seem to have similar strong wills.

Walter's face turned red and he got mad. Why the hell did Egon tell you that? Also, Erich was now pretty upset. He had believed that he and Walter were almost friends. But now Walter had bluntly lied to him. After this brief conflict, their relationship was never the same—they both had tension and had weakened trust in each other. No more joint ball games or anything like that.

AQVM

Egon: responsibility
Erich: truthfulness, respect, loyalty
Walter: honesty, respect, loyalty

Licenses

{L} Erich's psychological self-license: (1) they were almost friends and he trusted Walter; and (2) he wanted to talk about possible opportunities abroad and his future career. Apparently, they were not real friends, but just a superior and his close subordinate.

{M} Walter's moral self-license: (1) he was not as permissive to Erich of such questionable behavior as he had been to himself a few years earlier; (2) he was ashamed of his previous disloyalty; and (3) he felt that he had lost his face. He should have controlled himself better in this rather harmless situation.

{N} Egon's psychological self-license is not obvious: it is hard to see what "under special circumstances" could mean, but most likely Egon revealed such sensitive information only for some benefit of his company—certainly not as gossip due to his mature and fair personality. This remains partly an open question.

Analysis

The Utilitarian Perspective

If we judge Erich's request (or ultimatum) based on its consequences, then it seems like the wrong thing to have done. Trust was lost, feelings were hurt, and a friendship ended. In this light, Erich did the wrong thing by requesting a temporary assignment, and threatening to leave the company if it did not comply. But notice, this is not to suggest that a utilitarian would condemn *all* instances of extortion. Walter appears to have done much the same thing as Erich, and to positive effect. Walter was granted a two-year position in Japan, after which he continued working for his company. This indicates that the consequences of *his* request were good, and, thus, that his instance of extortion was not wrong. We cannot know for sure, as the case at hand does not include enough details of the fallout from Walter's past action. Suffice it to say, while there is nothing wrong with wanting to be temporarily assigned to a foreign subsidiary, there could be something wrong with the way a person acts on that desire.

The Deontological Perspective

One of the appealing aspects of Deontology is its insistence on consistency and fairness. If a rule is good, then it applies to everybody; and to consider oneself an exception to the rule would either be inconsistent or require plenty of logical justification. Thus, if Walter is in fact condemning Erich's request after having made a similar one himself, then a deontologist would find that problematic. Indeed, all acts of extortion, blackmail, and bribery would be considered wrong by the deontologist, regardless of whether or not they "work." Unsurprisingly, these ways of acting violate both versions of the Categorical Imperative. They involve treating someone as a means to an end, and they could not achieve goals in a world where they were universally practiced.

The Virtue Theory Perspective

Erich is not wrong to have ambition and a desire for the kinds of experiences that a temporary assignment at a foreign subsidiary would provide. However, he would have been wise to articulate the importance of such an opportunity in more diplomatic terms. Rather than issuing an ultimatum—"give me this opportunity or else…!"—Erich ought to have practiced some patience and magnanimity. He ought to have explained to Walter why he wanted this kind of assignment, offering reasons instead of threats. That approach would have been less extreme and confrontational.

For his part, Walter should have reacted more moderately to Erich's demand. Instead of accusing Erich of extortion, Walter should have responded a bit more charitably, chalking the demand up to the impetuousness of youth. Giving Erich the benefit of the doubt, Walter might have asked Erich why he felt so strongly, and suggested that a "softer" approach might be more productive. But instead of offering a young professional a lesson in practical wisdom, Walter reacted defensively and aggressively.

Discussion

Erich was no negotiator, as he immediately reinforced his reasonable desire with extortion. Therefore, his non-diplomatic behavior was similar to that of the other Erich in Box 7.3. They were both young and inexperienced professionals. Negotiation requires both skills to negotiate and patience to wait for the desired results [6]. Erich had neither of these qualities. His approach was blunt and just plain wrong. He should have realized that Walter was not his buddy, but instead his superior—even though they seemed to have a very good relationship. In general, it is wise to be (a little) sensitive when interacting with one's superior. Also, Erich should not have quoted the apparently confidential information about Walter that he had heard from Egon—it was thoughtless. And finally, he should have shown sincere remorse for the unfortunate conflict; his good relationship with Walter was certainly of mutual benefit, but their poor interaction destroyed it.

Walter was a benevolent and rational person—a mature adult. It was not usual for him to lose his temper and act irrationally, like he did after Erich first expressed his own extortion and later revealed the fact he heard from Egon about Walter's earlier extortion. But what could be the reason for Walter's unexpected behavior? Perhaps he was suffering from a long-term *ethical hangover* from his similar extortion act a few years ago. He was known to be a loyal employee; and Erich's extortion might have triggered a kind of anxiety response because it brought his (hidden) "shameful" behavior back to his mind (see Table 2.1 for possible symptoms of an ethical hangover). Of course, this is just speculation. Nonetheless, as an experienced department manager, he should have been able to "pull the brakes" to avoid such a "crash." And afterwards, Walter, as a "senior," should have been able to restore the inflamed relationship. But, unfortunately, that did not happen.

REPEATED VIOLATIONS OF ACADEMIC INTEGRITY

The relationship between a professor and his senior student, in a fairly small college, is somewhat like that between a superior and his subordinate in an engineering organization. Therefore, we decided to also include one vignette from the university world at the end of this interaction chapter. Because plagiarism, cheating, and academic dishonesty have gained considerable attention in the past decade, we chose a recurring case where the violation of academic integrity was obvious. And this, for sure, led to difficult interactions between the soon-to-be-graduating senior and his advising professor.

Never Again

BOX 7.6 VIGNETTE

ACTORS

Gunther: visiting professor
Kurt: electrical engineering senior
Heidi: computer engineering senior

CASE

Gunther was a visiting professor of electrical and computer engineering at a Research 2 (i.e., high research activity) American university. He had his 12-month sabbatical from a European institute. In addition to his scholarly research on nano-robotics, Gunther taught two senior-level courses: embedded real-time systems (fall semester) and real-time operating systems (spring semester). Both of these courses had about 20 students.

At the end of the fall semester, Gunther's teaching assistant told him that Kurt's and Heidi's personal project assignments had exactly the same program codes—even the comments between instructions were equal. That looked like an obvious integrity violation. Gunther recalled the university's Honor Pledge, which stated that "students uphold highest standards of honesty and integrity in all their academic work."

He decided to meet separately with Kurt and Heidi. When Kurt came to his office, Gunther asked: "Do you know why I wanted to meet you?" Kurt replied that he had no idea. Gunther showed him the program listing of Kurt's project assignment side-by-side on the listing of Heidi's program, and asked: "Why are these two programs exactly the same?" Kurt pretended surprised and answered: "I don't know." Gunther stayed calm and said that either you or Heidi or both have cheated; this was supposed to be a personal project—and you knew that.

After a serious and stressful conversation, Kurt eventually admitted that he had copied Heidi's program code and submitted it as his personal contribution. He also told that Heidi had nothing to do with this cheating issue. Kurt had copied the file to his USB-memory in the computer lab when Heidi went to her lunch and forgot to log out from her computer {O}. And as Heidi was a "straight-A" student, Kurt thought her program should definitely work.

Gunther was aware that if a student commits an academic violation, he may sanction the student, for example, by adjusting the student's grade. Kurt's final grade for the course was going to be B+, and Gunther reduced it moderately to C. Kurt said that the sanction was fair, and he promised he will never violate the Honor Pledge again. But Gunther somehow could not trust in Kurt's word.

So, at the end of the following spring semester, a similar case of cheating took place, in which Kurt was again involved. And Gunther asked Kurt to his office. When Kurt arrived, Gunther said: "Please tell me the reason why you are here." Kurt had no idea at all. This time, Gunther said that he could wait as long as Kurt tells the reason himself. Time passed while Gunther was writing an e-mail; Kurt sat next to him quietly. After about 20 minutes, Kurt finally opened his mouth and said: "Maybe you wanted to meet me and discuss about the course project." Yes, that was the reason!

Now, Gunther had trouble remaining calm, he was getting angry. He said: "It's only four months ago when you promised me that you will never violate the Honor Pledge again." Then, they had another difficult discussion that ended up to a sanction: now Kurt's grade was reduced more significantly, from A– to C– {P}.

Gunther stood up and told Kurt that in a couple of weeks he would return to his home institute and they would likely never see again. He also said firmly: "Wake up, Kurt, and try to understand that you can't build and maintain your engineer's career on the basis of plagiarism—nobody can; thus, for your own sake, learn to behave ethically!" Kurt was embarrassed and he timidly replied: "Yes, sir." Gunther was truly worried about Kurt.

At the end of May, Kurt graduated with a bachelor's degree in electrical engineering with a fairly average GPA. Kurt's dad and grandpa were also electrical engineers; they were proud of their boy.

AQVM

Gunther: responsibility 🙂
Kurt: self-discipline 🙁, integrity 🙁, respect 🙁
Heidi: responsibility 🙂

Licenses

{O} Kurt's moral self-license: (1) he had difficulties with his own program code; (2) he got a perfect opportunity to cheat—only this one time; and (3) the visiting professor would not pay attention to such matters, as he was mainly interested in his research.

{P} This time, Kurt's moral self-license had four elements: (1) he had overwhelming difficulties with his program code for the spring project; (2) he absolutely wanted to graduate at the end of this semester; (3) since he was caught already in December, he could not be so unlucky as to be caught again; and (4) he considered the previous sanction to be tolerable. So, obviously, Gunther should have done something else after the first cheating incident to prevent this second one—but what constructive could he have done?

Analysis

The Utilitarian Perspective

Kurt cheated; twice (that we know of). In the short term, this caused disappointment and frustration on Gunther's part, embarrassment and anxiety on Kurt's. But in the long term, is this really a problem? Gunther had a career to continue and Kurt, having graduated, had a career to commence. After this small blip on the radar, everyone moved on with their lives relatively unaffected. If consequences are really all that matter, then cheating is not wrong in and of itself. It is only wrong when it brings about less happiness than it might have. So, from a utilitarian perspective, those cheaters who are never discovered (and who are pleased by their having cheated) have not done anything wrong. Once again, this ought to raise serious questions about the utilitarian moral framework.

Those of us analyzing this case objectively can only hope that Kurt either heeded Gunther's advice and changed his ways or, failing that, was never put in a position that could involve negatively affecting others.

The Deontological Perspective

Kurt cheated, and he did so in order to earn a high mark on an assignment. Does this action pass the test of the Categorical Imperative? Probably not. The maxim "When you want to do well on a difficult assignment, cheat" is not universalizable. This much is made evident by the two instances of cheating discussed in the case. After the first, Gunther was already suspicious of Kurt. Imagine a world in which everybody cheated, and you will see how that course of action would not allow one to achieve their goal of doing well on an assignment. Cheating does not stand in light of the PU.

What about the PH? By cheating, Kurt disrespects Heidi by treating her as a means to an end. He disrespects Gunther by ignoring his good and worthy advice against cheating. And if he believes that cheating is fine for him but not for anybody else, then Kurt disrespects all of the other students in his class by supposing that the rules do not apply to him. If he believes that it is fine for everybody to cheat, then the PU will show the faulty reasoning behind that belief. From a deontological perspective, Kurt acts wrongly and is motivated wrongly.

The Virtue Theory Perspective

Kurt is a cheater. Or, if that is too hasty an assessment, then at least Kurt is on his way to becoming a cheater. The actions you perform and the habits of character you cultivate determine who you are as a person and, by extension, as a professional. Persons who cheat once might not be acting from a vicious character. They may be "testing the water," as it were. But deciding to cheat even once reveals poor judgment and weak character. And the more one cheats, the more likely one is to become a cheater.

A person of integrity would not succumb to the temptation to cheat. Having integrity means possessing the character traits, the experience, and the insight to make the right judgments and decisions. Virtues are integral to goodness of character in much the same way that a spark plug is an integral part of a traditional internal combustion engine. Without the spark plug, the engine would not work the way it is supposed to. It will still be an engine—just not a good one. And without virtues, a person cannot function well as a person. Someone with a weak will and a vicious character is still a human, but not a good one. Having integrity as a person, then, means having all the "pieces" that are necessary for living excellently in a community of others. And when a person has integrity, he/she can not only become a good engineer—but also an engineer who is good.

Discussion

During his year-long sabbatical, Gunther looked forward to fruitful research collaboration and stimulating undergraduate teaching of his favorite lecture topics. Hence, he felt quite uncomfortable with the recurring cheating incidents. Nonetheless, Gunther tried to guide Kurt away from the unethical pathway. Maybe he should have discussed with the department chair about handling those two integrity violations; but Gunther did not want to make them *official* issues. He was rather constructive when dealing with Kurt and deciding on the deserved penalties—he had seen similar behavior a few times at his home institute, as well. But eventually he could not trust Kurt anymore.

Kurt bluntly acted against the Honor Pledge of the university. (What is the *concrete* value of such pledges, are they merely ethics-washing?) When Kurt copied Heidi's program file behind her back to his USB memory stick, "opportunity made the thief"—as discussed in Chapter 4. And he was apparently satisfied with the moderately reduced final grade. But in the spring semester, Kurt took the risky and wrong path again; and he was caught again. He should definitely stop cheating until he becomes a habitual cheater. After these two cheating occasions, other can come more easily. As Sissela Bok stated in her book: "Psychological barriers wear down; lies seem more necessary, less reprehensible; the ability to make moral distinctions can coarsen; the liar's perception of his chances of being caught may warp" [2] (p. 25). This is also true of cheating in all its forms. However, Kurt's facade toward his dad and grandpa remained clean; they saw him as their good boy. Now, after several years, Kurt is a professional engineer and thus a kind of role model for the newly graduated engineers in his company—hopefully an ethical role model.

SUMMARY

Interpersonal interaction is one form of professional communication, and communication skills are considered an important asset in an engineer's toolbox [7]. The personal wellbeing of engineers and the effectiveness of their organization largely depend on the success of the interaction between cooperating individuals. Even fairly minor distortions in day-to-day interactions can eventually lead to significant consequences related to an individual's career development, organizational productivity, the appropriateness of a chosen technology, and so forth.

In this chapter, all the six case studies have an interaction connection. Table 7.1 shows a summary of the main reasons behind the related psychological/moral self-licenses. Here, half of the licenses were granted for the "Company's benefit." This is about the same proportion as with the product-related cases. And approximately one-fifth of the licenses were for the "Actor's benefit," while with the product cases that proportion was one-third. On the other hand, the corresponding figures with employment-related cases were notably different; except that the proportion of "Actor's benefit" licenses was roughly the same.

Further, from Table 7.2, we can see that the virtues of *respect* and *responsibility*, as well as the meta-virtue of *honesty–integrity–truthfulness* were violated the most, while self-discipline was violated only twice. Intuitively, it is understandable that there will be problems in person-to-person interaction if there is no/poor respect between the interacting individuals.

In Chapters 5–7, we presented 20 micro-level case studies in depth. These studies are unique compositions because they deal with the unethical cases from the viewpoints of behavioral ethics (Licenses) and moral theories (Analysis). In addition, the Discussion sections provide a complementary commonsense perspective. As we mentioned already in Chapter 1, Kim et al. stated after an academic–industry workshop on engineering ethics [8]: "It became clear that every engineering decision or project raises ethical issues and that engineering students need to understand this day-to-day aspect." And Chapters 5–7 form a solid response to this important

TABLE 7.1
Main Reasons behind the Speculated Psychological/ Moral Self-Licenses {A} to {P}

Self-license	Company's Benefit	Actor's Benefit	Other
A	Yes		
B	Yes		
C	Yes		
D			Yes
E			Yes
F	Yes		
G	Yes		
H		Yes	
I	Yes		
J			Yes
K	Yes		
L			Yes
M			Yes
N	Yes		
O		Yes	
P		Yes	
%	**50**	**19**	**31**

educational challenge. We hope that the reader—whether a student or a young professional—now has a clearer understanding of the various ethical issues that engineers may face in their *daily* work life.

Next, Chapter 8 presents a few macro-level cases and discusses their short- and long-term consequences, which fall into the category of disasters, scandals, as well as environmental issues.

TABLE 7.2
Actor's Qualitative Virtue Measures in Interaction-Related Cases

Virtue	Strength Level 🙂	Strength Level 😐	Strength Level 🙁
Determination	1	0	0
Fairness	1	1	1
Honesty	0	0	**3**
Integrity	0	0	**3**
Loyalty	2	1	1
Respect	0	4	6
Responsibility	3	1	5
Self-discipline	0	0	2
Tact	0	0	1
Truthfulness	1	1	**1**

CLASS EXERCISES

1. Is it ethically right to serve alcohol (wine and beer) or even make it freely available to employees at company events, such as on the division's recreation day in Box 7.2?

   Class debate: The instructor moderates a prepared debate over Question 1. First, the class is divided into three teams; one of the teams defends the positive answer ("yes"), one team supports the negative answer ("no"), and the third team evaluates the arguments presented. What are the objective conclusions of the evaluation team?
2. How could the long-lasting rancor between Egon and Gunther in Box 7.3 be relieved or even eliminated, as it adversely affects the effectiveness of their collaboration on joint projects? It would certainly be to their company's benefit. You can use an AI chatbot as your assistant.
3. Edit the text of the vignette in Box 7.4 to be ethical. Is the end result such that it could have come true in practice? Justify your answer.
4. Form an acting pair of two volunteer students. They first prepare a script for a role-play based on the vignette "This Is Extortion!" (Box 7.5). And then they perform it in front of the class.

   Class discussion: Consider the underlying case and try to find a plausible explanation for Walter's unexpected hostile behavior.
5. Was Erich's behavior in Box 7.1 right by your own moral standards, or how should he have acted in such a stressful legal interview? If you proposed an alternative course of action, analyze its consequences for all the stakeholders.
6. In Box 7.6, Gunther did not report Kurt's integrity violations to the university, even though that was the recommended practice; he just lowered the final grades moderately/considerably. Did Gunther do the right thing, or how should he have acted in this case? Administration viewpoint versus Kurt's viewpoint.
7. Which one of the six unethical cases in this chapter had the most serious consequences? Why do you think so? See Table 4.3 for specifics of consequences.

   Class discussion: Discuss the answers of different students and try to form a common opinion of the whole class.
8. Which *single* virtue from Table 1.1 would you give to all the actors in Boxes 7.1–7.6 to make their behavior more ethical in their corresponding cases? Justify your answer.

REFERENCES

1. A. Brei, "Being loyal and being ethical," *IEEE Potentials*, vol. 41, no. 3, pp. 46–48, 2022, doi: 10.1109/MPOT.2020.2989712
2. S. Bok, *Lying: Moral Choice in Public and Private Life*. New York, NY: Pantheon Books, 1978.
3. Á. L. de Nalda, M. Guillén, and I. G. Pechuán, "The influence of ability, benevolence, and integrity in trust between managers and subordinates: The role of ethical reasoning," *Business Ethics: A European Review*, vol. 25, no. 4, pp. 556–576, 2016, doi: 10.1111/beer.12117

4. R. Cropanzano and M. S. Mitchell, "Social exchange theory: An interdisciplinary review," *Journal of Management*, vol. 31, no. 6, pp. 874–900, 2005, doi: 10.1177/0149206305279602
5. D. Tjosvold, "Power and social context in superior-subordinate interaction," *Organizational Behavior and Human Decision Processes*, vol. 35, no. 3, pp. 281–293, 1985, doi: 10.1016/0749-5978(85)90025-1
6. M. H. Bazerman, J. R. Curhan, D. A. Moore, and K. L. Valley, "Negotiation," *Annual Review of Psychology*, vol. 51, pp. 279–314, Feb. 2000, doi: 10.1146/annurev.psych.51.1.279
7. S. J. Ovaska, "Managing your career in a dynamic environment," *IEEE Potentials*, vol. 37, no. 3, pp. 24–26, 2018, doi: 10.1109/MPOT.2017.2764512
8. D. Kim, B. K. Jesiek, C. B. Zoltowski, M. C. Loui, and A. O. Brightman, "An academic-industry partnership for preparing the next generation of ethical engineers for professional practice," *Advances in Engineering Education*, vol. 8, no. 3, 2020. [Online]. Available: https://advances.asee.org/wp-content/uploads/vol08/issue3/Papers/AEE-NAE-Brightman.pdf

8 Macro-Level Cases

Learning Objectives

After studying Chapter 8, you will be able to

- Recognize the potentially disastrous and scandalous nature of macro-level unethical behavior
- Identify the negative and positive consequences of macro-level unethical actions, taking into account different perspectives of the case
- Understand the human rights viewpoint to environmental ethics
- Realize that environmental ethics dilemmas can be highly diverse and cumbersome

In this chapter, we switch from micro- to macro-level unethical cases. While the macro-level cases form the focus in several other books on engineering ethics, our main focus is on the micro-level; and we draw on macro-level examples principally to demonstrate the disastrous or scandalous nature that some of these cases possess. *The Pocket Oxford Dictionary* defines *disaster* as "sudden or great misfortune; complete failure." And *scandal* is "(thing that causes) general feeling of outrage or indignation." We chose the Chernobyl nuclear disaster in the Ukrainian Soviet Socialist Republic (SSR) as the first example and the Volkswagen diesel emissions scandal as the second. These particular cases were chosen for their exceptionally shocking nature.

In addition, we discuss environmental ethics and a few macro-level ethical dilemmas related to it. It is worth noting that even the above-mentioned Chernobyl and Volkswagen cases are actually environmental ethics incidents because they caused more or less environmental pollution: nuclear radiation and toxic oxides of nitrogen (NO_X), respectively. And both of these pollution types—although different in many ways—are of serious health concern.

CHERNOBYL, 26 APRIL 1986

The following presentation is *not* intended to be another review of this well-documented case. Instead, our aim is to show the complexity and multidimensionality of the rather unusual chain of events over a lengthy period of time. And to deal with the multidimensionality, we take different perspectives on the case as well as its consequences. In this way, we hope to provide a fertile basis for class discussions about the role of unethical practices behind the vast disaster at the Vladimir Ilyich Lenin Nuclear Power Plant.

Violations Behind

As summarized by the World Nuclear Association (WNA) [1], the Chernobyl disaster on Saturday, 26 April 1986 was the result of a flawed nuclear reactor design,

DOI: 10.1201/9781003485520-8

and the 1,000 MW Reactor No. 4 in question was operated by insufficiently trained personnel. Baneful mistakes made by the plant operators caused two back-to-back steam explosions and subsequent fires, which released roughly 5% of the radioactive reactor core into the environment and spread radioisotopes (such as I-131, half-life 8 days; Cs-137, half-life 30 years; and Sr-90, half-life 29 years) to many parts of Europe—even to distant Finland, which is located more than 1,000 km north of Chernobyl. Ironically, the disastrous explosions took place during an experiment testing a procedure to cool the reactor core in an *emergency situation*. The ongoing test was preceded by a series of operator actions, including the disabling of automatic shutdown mechanisms. This catastrophe can also be seen as a consequence of the Cold War isolation of the Soviet Bloc and the resulting lack of appropriate safety culture. For additional information, see references [1, 2].

The condensed description above contains four ethical concerns that deserve our attention:

1. Flawed design of the nuclear reactor and all reactors of the same type (*engineering failures*)
2. Insufficiently trained personnel (*management problems*)
3. Serious mistakes by the plant operators (*operators' negligence*)
4. Lack of appropriate safety culture (*general attitude issue*)

All of these concerns involve violations of the ethical value *responsibility*, leading to obvious violations of *fairness*, *integrity*, and *respect*. The design of safety-critical components and systems—such as reactors for nuclear power plants—must be based on thorough uncertainty and risk management, as well as error-free calculations and system simulations performed by skilled engineers with utmost care. Besides, all the materials used and the construction/inspection/certification work must be of high quality; nothing is allowed to slip. Unfortunately, this was not entirely the case with the destroyed nuclear reactor. And it can be speculated that the fourth ethical concern was actually the principal cause behind the other three concerns.

If we apply the *extended licensing* principle proposed in Chapter 4 to this case, we can conclude that the Chernobyl disaster was not a pure accident but a stochastically licensed outcome—mainly because of the flawed reactor design. It was, therefore, not the result of a simple unethical act according to the straightforward model of Figure 2.1 but a multidimensional set of consequences that had gradually developed over the years due to the ethical concerns 1–4 above (and possibly others). And on 26 April 1986, at 1:23 a.m., the negligence of the plant operators on duty was the final trigger that led to the disaster.

Different Consequences

Immediate Consequences

What does "immediate consequences" mean in this particular case? Arguably, the case actually began to emerge on the day Reactor No. 4 was commissioned in 1983, initiating the four *ethical concerns* mentioned above. And one might have even expected that something rather serious would inevitably happen sooner or later with any reactor of the same type. Maybe something mildly or moderately serious had already happened before 1986, but it did not reach the Western news headlines—who could have known? Such suspicion is perhaps warranted due to the *secrecy* between East and West during the Cold War, especially when it came to nuclear matters. Today, we know that there had, indeed, been moderate problems already before 1986. And any secrecy of issues related to public safety is a form of dishonesty.

Interestingly, we can see an analogy between the Chernobyl disaster and a fatal myocardial infarction (heart attack), which is typically the result of coronary artery disease (heart disease). Such a heart disease (or the aforementioned set of *ethical concerns*) develops over years, and eventually when necessary conditions are met, it causes a heart attack (or the *two steam explosions and subsequent fires*). Hence, Reactor No. 4 and its stakeholders had been "sick" already for a few years.

Short-Term Consequences

The following hard data are mainly taken from the WNA report of reference [1], which summarizes the outcomes of the disaster. The explosions killed two operators of Reactor No. 4, and 28 operators and firemen died within a couple of months as a result of acute radiation syndrome. Nobody offsite suffered from *acute* radiation effects. The destructive fire burned for 10 days [2].

Large areas of the Belarusian, Ukrainian, and Russian SSRs and beyond were contaminated to varying degrees by radioactive fallout. At the time of the explosions, there were 115,000–135,000 inhabitants within a radius of 30 km from the Chernobyl power plant. But altogether 350,000 people had to be evacuated from their homes as a result of the disaster.

During the first week after the two explosions, the wind blew over the burning reactor—from the Black Sea toward northwest—and carried radioactive clouds over Finland, Sweden, and Norway. Since it rained and snowed intermittently on these days, radioisotopes entered the ground and thus caused damage to the agricultural sector in some areas. Due to the general secrecy of the Soviet Union, the first elevated radiation levels *outside* the Soviet Block were detected in Southern Sweden—one and a half days after the catastrophic event. Before this, the Western world had no idea of any nuclear accident. Almost three days after the explosions, Moscow finally released its brief official statement. Then, on the first week of May, the wind began to blow to the west, moving the radioactive clouds toward major Central European cities, such as Warsaw, Berlin, Prague, and Munich—and beyond. Fortunately, the radioactive contamination of the clouds had already decreased to some extent.

A sorrowful consequence of the misunderstandings surrounding this nuclear disaster was that some physicians advised pregnant women to have abortions due to radiation exposure, even though the levels in question were far below the likely teratogenic effects.

In many European countries, the Chernobyl disaster halted the planning and construction of nuclear power plants for years. The nuclear power industry was suffering, and nuclear engineering programs at several universities were scaled down or focused on safety issues and waste management.

Long-Term Consequences

Overall, the disaster is reported to result in "thousands of deaths," including a large number of people suffering from premature cancers linked to exposure to the deadly radiation [2]. On the other hand, the United Nations (UN) Scientific Committee on the Effects of Atomic Radiation concluded in 2006 that, in addition to approximately 5,000 cases of thyroid cancer—resulting in 15 deaths—"there is no evidence of a major public health impact from radiation exposure 20 years after the accident" [1].

Today, radioactive cesium can still be detected, for instance, in natural berries and mushrooms—which people eat—in some areas of Finland and Sweden, albeit in fairly small quantities. But a significant proportion of wild boars hunted in Bavaria, Germany, cannot be eaten due to their unhealthy levels of Cs-137 [3]. In 2018, a friend of Ovaska's visited Chernobyl and the surrounding area; what he saw was miserable: an abandoned amusement park in the Ukrainian *ghost town* of Pripyat (about 3 km from the Chernobyl power plant) is a vivid reminder of the long-awaited summer fun of 1986 that never came (Figure 8.1).

Finally, as a *positive* consequence, the Chernobyl disaster led to major improvements in the nuclear safety culture within the Soviet Block and to constructive East–West nuclear cooperation. It caused a kind of *ethical hangover* that gave stakeholders a valuable chance to consider. As we stated in Chapter 2, "every ethical hangover is a chance."

FIGURE 8.1 A quintet of rusting bumper cars in the abandoned amusement park in Pripyat, once a bustling "atomic city" of nearly 50,000 residents. (Photo (8 September 2018) courtesy of Kullervo Hakama.)

In 2006, former President Mikhail Gorbachev made a thought-provoking statement in an interview: "Even more than my launch of perestroika, [Chernobyl] was perhaps the real cause of the collapse of the Soviet Union five years later" [4]. (*Perestroika* was his liberal reform program.)

At the moment, when the world is in the middle of the dilemma of fossil fuels and energy production, nuclear power is seen (again) as a "clean" alternative in many countries. However, there are exceptions to this changed opinion: Germany, for example, closed its *last* three nuclear power plants in 2023.

VOLKSWAGEN "DIESELGATE"

We already introduced the "Dieselgate" in "The Diesel Emissions Scandal" section of Chapter 1. So, in the present chapter, we focus on the *consequences* of this case publicized on 18 September 2015. (Incidentally, the suffix "-gate" refers to the infamous Watergate scandal in the 1970s.)

VIOLATIONS BEHIND

Daniel ("Dan") Carder and his small team of researchers at the West Virginia University's Center for Alternative Fuels, Engines, and Emissions discovered that Volkswagen had been installing a *defeat device* in its diesel engines. "The emissions we were measuring in the real world over the road were appreciably higher than... when we took them to the California [Air] Resources Board's vehicle emissions testing lab," Carder said in an interview with the German news broadcaster Deutsche Welle shortly after the striking discovery [5]. And he continued: "The emissions levels there were significantly lower than what we saw in the field." Hence, the malicious software enabled these vehicles to *pass* emissions tests under laboratory conditions while emitting up to 40 times the level of nitrogen oxides allowed in the United States during normal use.

The term nitrogen oxides (NO_X) describes a mixture of nitric oxide (NO) and nitrogen dioxide (NO_2), which are highly reactive gases that contribute to ground-level ozone formation through reactions with volatile organic compounds in the presence of sunlight.

The infamous software "patch" was installed in 11 million diesel vehicles worldwide, including 590,000 automobiles in the United States. Brands involved were: Volkswagen, Audi, Porsche, Škoda, Seat, and the Volkswagen Commercial Vehicles, which all used the turbocharged EA189 diesel engine. Eventually, Volkswagen pleaded guilty to fraud, obstruction of justice, and falsifying statements.

Dan Carder can be seen as an accidental whistleblower. He was just working under an ordinary testing and measurement contract for the International Council on Clean Transportation when the discovery was made—it appeared as an unexpected side result. For this outstanding act, Carder was named to the 2016 *Time* 100—the magazine's annual list of the 100 most influential people in the world. His achievement brought significant recognition to the West Virginia University community, underscored by their spirited cheer, "Let's Go Mountaineers!"

Why didn't anyone within the Volkswagen Group blow the whistle earlier?—Well, all of the involved persons were obviously loyal employees and did what they were told. Of course, anyone who would have stood up to object would likely have been swept aside. But loyalty, a virtue, becomes a *vice* when pushed to extremes; and even altruistic participation in deceptive behavior is definitely something that cannot be *demanded* of a loyal employee. This virtue–vice issue was covered in Chapters 1 and 3.

Different Consequences

Rather than discussing the consequences of this Volkswagen case from the *temporal* perspective, as in the Chernobyl case (i.e., immediate, short, and long term), it is more instructive to look at the *spatial* consequences for the different stakeholders of the case, i.e., Volkswagen Group's shareholders, the company itself, customers, and others. In this case, too, we can identify both negative and positive consequences of unethical behavior, as we did in the case of Chernobyl.

Shareholders

The scandal must have been a tremendous shock to the shareholders of the Volkswagen Group, as the price of the ordinary share on the Frankfurt Stock Exchange fell by about 41% in the first month after the scandal was announced. On 17 September 2015—the day before the scandal became public—the closing price was €167.80, while on 19 October 2015, it had fallen to €99.19. For comparison, the price of the ordinary share was €118.45 at the end of 2023 and €107.15 on 11 July 2024 (*Source*: finance.yahoo.com).

Based on this cursory analysis, it appears that the share price is still—nine years later—at the reduced level at which it fell in 2015 (naturally, it has fluctuated over the years). This is apparently related to low *investor confidence* in the Volkswagen Group. In addition, the market sentiment is not necessarily favorable. Or could it be that the entire automotive industry is suffering? This is not the case, as Toyota, Volkswagen's biggest competitor, seems to be doing pretty well on the New York Stock Exchange. At the end of 2015, Toyota Motor Company's share price was $116.50, but on 11 July 2024, it was $206.45, representing a 77% increase. Surprisingly, Toyota makes about the same number of automobiles per year as Volkswagen but with roughly *half* the number of employees.

Although the share price is not doing well, the dividend paid to shareholders has risen nicely since 2015 (see Table 8.1). The years 2016–2018 showed significant growth, which was saturated in 2019 and 2020. But the year 2021 was the beginning of a new upward trend, which, however, seems to be slowing down. If we take into account the Volkswagen share price at the end of 2023 (€118.45) and the dividend per ordinary share for 2023 (€9.00), we get a 7.6% return on invested capital, which is a good result.

All in all, the shareholders of the Volkswagen Group suffered significantly from the diesel emissions scandal. And that suffering is only partially over; although the dividend per share is already at a good level, the share price still does not show a satisfactory development.

TABLE 8.1
Dividends Paid by the Volkswagen Group for the Fiscal Years 2013–2023

Fiscal Year	Dividend Paid in Billions of €	Change from Previous Year in %
2013	1.9	n/a
2014	2.3	+21
2015	**0.068**	**–97**
2016	1.0	+1370
2017	2.0	+50
2018	2.4	+20
2019	2.4	0
2020	2.4	0
2021	3.8	+58
2022	4.4	+16
2023	4.5	+2

Source: Annual reports of the Volkswagen Group.

At this point, it is interesting to take a look at the voting rights distribution of Volkswagen's shareholders on 31 December 2023:

- Porsche Automobil Holding SE, 53.3%
- State of Lower Saxony, 20.0%
- Quatar Holding, 17.0%
- Free Float, 9.7%

Hence, Porsche Automobil Holding SE holds a dominating portion of the voting rights in the Volkswagen Group. However, the state of Lower Saxony, which has a 20% voting share, can veto important decisions, including any remedial reductions of the workforce within its state. Such a biased distribution of voting power can slow down the renewal of the Volkswagen Group; and investors may perceive it negatively. Volkswagen is indeed a massive "ocean liner," and changing its course requires vision, strategy, effort, time, and, of course, consensus among the key players.

Company

As a corrective measure—after the EPA officially disclosed the misconduct—Volkswagen implemented a *lean* organizational structure that would facilitate direct communication with top management and increase efficiency [6]. We believe that during the fraud years, the company's top management practiced a strictly authoritarian leadership style.

As expected, the CEO stepped down; Matthias Müller from Porsche replaced Martin Winterkorn as CEO in September 2015. But already in April 2018, Herbert Diess, who had joined VW from BMW in 2015, succeeded Müller. And, in September 2022, Oliver Blume, also the CEO of Porsche, took over the position. Hence, being the CEO of the Volkswagen Group in the post-"Dieselgate" era appears to be a fairly turbulent role. In all, about half of the senior managers were fired, and several managers have faced legal consequences—one was sentenced to seven years in prison. Even the former CEO (January 2007–September 2015), Martin Winterkorn, has finally gone on trial in Germany, nine years after the diesel emissions scandal was revealed. He faces charges of perjury, market manipulation, and fraud.

It is estimated that Volkswagen spent more than $30 billion on various settlements and related expenses. This huge loss is roughly the same as the dividends paid by the Volkswagen Group to its shareholders for the 11 fiscal years 2013–2023, in total. Understandably, the company management and labor unions agreed to reduce the number of employees—by 30,000 and mostly in Germany—to cut costs [6]. And all of this was the result of unethical business practices *licensed* by high-level management. Moreover, at the same moment that the unethical "go ahead" command was given to the personnel involved, a number of stochastic consequences were also licensed under the *extended licensing* principle—and many of them were indeed materialized, as we see in this section. Think about that.

Volkswagen's own assessment of the causes of the scandal included three factors [7]:

1. Misconduct and deficiencies of individual employees
2. Weaknesses in some processes
3. A mindset in some areas of the company that tolerated breaking the rules

In one form or another, these factors are behind just about every corporate fraud crisis.

According to the recently released TRANSFORM 2025+ strategy [8], Volkswagen aims to realign itself. By 2030, they want to increase the share of fully electric cars in Europe to 70% of their sales. And their ambitious goal is to make the *Volkswagen brand* the global market leader in climate-neutral electric mobility. In 2023, the whole Volkswagen Group delivered over 771,000 fully electric vehicles worldwide. To put this into perspective, the corresponding number of Tesla electric cars was about 1.81 million. Most obviously, the era of diesel cars is waning. These are *positive* consequences of the emissions scandal, which not only speeded up Volkswagen's own transition to electric vehicles but also accelerated the electric car strategies of other automakers in Europe, China, and the United States.

As a gratifying signal, the Volkswagen Group's 2024 code of conduct, *Our Code*, promises that "We are honest and speak up when something is wrong" [9]. Furthermore, they state: "Integrity acts as an inner compass to do the right thing out of our own conviction—regardless of economic or social pressure." Yep, that's the way to go!

Customers

Volkswagen agreed to expensive settlements with their customers in a few countries, notably in the United States, but refused to do so in most others. In general, they have not done much to restore the *confidence* of customers. Studies have advised corporations dealing with a (self-inflicted) crisis to recognize the seriousness of the crisis and take responsibility for it, in order to improve their public image. Although Volkswagen's management did apologize, they did not take responsibility or try to explain the abuses [6].

With the short attention span of the public and the media, "Dieselgate" is fading from the minds of current Volkswagen owners as well as potential Volkswagen car buyers. Jung and Sharon pointed out in their aftermath overview [6]: "It appears that most consumers were more upset about the effect of the defeat device on their car's performance and the hassle of the recall than they were about the emission of poisonous gases into the environment—the true impact of which was difficult for most people to fully appreciate." Could this reflect a bias in the personal values of some customers?

On the other hand, automobile emissions issues are now better known and understood among car owners at large than before the scandal. Hopefully, this enhanced awareness will gradually guide car buyers to vehicles with lower emissions, both NO_X and CO_2.

Others

Since the scandal became public, car emissions standards and related regulations, as well as corresponding testing methods and certification procedures, have been developed in many countries. Even though the measured levels of nitrogen oxides were *illegal* in the United States and a few other countries, that was not the case in most European, Asian, and South American countries. Today, the situation has improved in these respects; the Volkswagen case woke up the authorities in many countries.

ENVIRONMENTAL ETHICS

Occasionally, engineers are responsible for developing technologies that may cause environmental damage; and they likewise look for innovative solutions to problems caused by such technologies. An internal combustion engine (which produces dangerous carbon monoxide) and a catalytic converter (which, among other functions, converts carbon monoxide to less hazardous carbon dioxide) are an example of an early innovation pair. But, as we know, carbon dioxide is a problematic gas in the atmosphere, necessitating further innovation from automotive and other engineers. And these innovations can lead to new environmental challenges, and so on. Such an evolving problem–solution chain keeps engineers busy.

Determining the *moral standing* of nature is the basis of discussing the ethical issues of environmentalism. Hence, ethics for engineers is not only about avoiding dishonesty, disloyalty, disrespect, and so forth between individuals/organizations. Environmental (including *space*) aspects are considered more and more important and are now included in professional ethical codes. The comprehensive *Code of Ethics for Engineers* of the National Society of Professional Engineers (NSPE) states: "Engineers are encouraged to adhere to the principles of sustainable development in order to protect the environment for future generations" [10]. And it further particularizes, "sustainable development is the challenge of meeting human needs for natural resources, ..., and effective waste management while conserving and protecting environmental quality and the natural resource base essential for future development." But it is surprising to observe that this is only in the form of an *encouragement*.

In this section, we first introduce a human rights viewpoint to environmental ethics. Three ethical dilemmas are then presented that fall primarily within the fields of materials/mechanical, space, and civil/electrical engineering. We deliberately chose very different case studies to illustrate the enormous diversity of environmental ethics issues involving engineers. Next, these dilemmas and environmental ethics in general [11] are addressed from the perspective of an individual engineer. Such dilemmas are generally approached from a *cost* (to the environment)–*benefit* (to society) standpoint, which is clearly utilitarian. However, it may take even decades after a disputed decision before the true cost is known and internalized. This makes these dilemmas difficult to handle.

A Human Rights Viewpoint

The following introductory discussion is adapted from Brei's dissertation in philosophy [12].

There are, in recent environmental literature, instances of tackling environmental issues via human rights. But is such an approach advisable? And what advantages does a human rights approach to environmental ethics have to offer? These are relevant questions.

First, there is a nice symmetry of scopes. Environmental problems are ubiquitous—there are precious few places on Earth where some environmental problem or other is not an issue. And some of the big environmental problems (like global warming) have the potential to affect everybody. Human rights, because they are rights one bears simply by virtue of being human, apply universally. Thus, a human rights approach seems to be an effective way to address issues that affect all humans.

Second, notice that both environmental problems and human rights transcend political boundaries. Attempts at dealing with environmental problems on a local, state, or even national level can be ineffective when such problems occur across borders. Dealing with these problems by way of human rights makes it less likely that differences between state laws will stall progress on the environmental front.

Third, human rights (in principle, anyway) correspond to our most stringent legal and moral imperatives, as well as to our most cherished and valued norms. Environmental issues are among the most pressing we face today. It is therefore reasonable to believe that among the imperatives and norms reflected by human rights ought to be those involving our relations with and effects on the natural environment.

Fourth, a human rights approach to environmental problems is a good way to get people concerned and involved. Human rights are more likely to be felt personally than are, for example, arguments and data from ecologists. Thus, people are more likely to be motivated to address environmental issues when these issues are put into human rights terms.

Fifth, to address environmental problems via human rights is to treat these issues with the appropriate degree of significance. To take this approach is to recognize (as we should) the importance of our natural environment and the problems that threaten it.

Sixth, and finally, approaching environmental issues via human rights allows one to build on the common ground occupied by environmentalists and human rights advocates. Furthermore, there are basically three views to consider in this human rights framework, detailed in reference [12]:

1. Our duties regarding nature stem from the human right to a healthful environment.
2. Our duties to nature arise from the rights *of* nature.
3. Our obligations regarding nature arise from the human right to health.

Generally speaking, a *right* offers us some standard for acceptable conduct. Rights tell us how we ought to behave toward one another and what we may expect from one another, particularly if we are interested in not harming one another. For instance, there is a duty not to disrupt or degrade the ecosystem to such an extent that health would be more difficult to attain. This is so because of the close connection between human health and the natural environment. According to the preamble to the constitution of the World Health Organization (WHO), health is a state of complete *physical*, *mental*, and *social* wellbeing and not merely the absence of disease or infirmity.

The *IEEE Code of Ethics* (see Appendix B) includes this aspect in connection to the environment: "[To] hold paramount the safety, health, and welfare of the public, to strive to comply with ethical design and sustainable development practices, …, and to disclose promptly factors that might endanger the public or the environment."

We chose to introduce the lesser-known human rights viewpoint to environmental ethics to complement our favored individual-engineer point of view in dealing with professional ethical issues. And, it is good to remember that the right to work is also a human right, included (among many other places) in the United Nations Declaration of Human Rights (Article 23).

Difficult Dilemmas

Wind Turbine Blades

Wind power is one of the key technologies of renewable energy. According to a report by Grand View Research [13], the global wind power market size was estimated to be close to $100 billion in 2021, and it is expected to grow at a compound annual growth rate of 6.5% during the period 2022–2030. If this is an accurate prediction, the market size will be about $175 billion in 2030.

Wind turbines intended for utility-scale applications are located in onshore and offshore wind farms connected to the nation's power transmission grid. Figure 8.2 shows an example of a typical wind turbine in a Norwegian wind farm, which consists of 15 turbines operated by Finnmark Kraft AS (finnmarkkraft.no/prosjekter/hamnefjell-vindkraftverk). Can you see the "Lilliputian" man standing next to the turbine tower?

FIGURE 8.2 A wind turbine at the Hamnefjell wind farm in Båtsfjord, Northern Norway; the height of the tower is 84 m, and the diameter of the rotor is 112 m, with the rated turbine power of 3.45 MW. (Photo (13 August 2022) courtesy of Helena Ovaska.)

In Figure 8.2, the length of a turbine blade is 54.7 m, and it weighs approximately 12,300 kg. Depending on the dynamic wind and weather conditions, the blades have a projected lifespan of 20–25 years. Furthermore, it can be estimated that there are currently several hundreds of thousands of utility-scale wind turbines in use; hence, their number of blades must be around a million.

However, wind turbine blades are a recycling challenge ("nightmare") due to the materials used and their complex composition [14]. The blades are typically made of fiberglass-reinforced plastics composites, which have considerable recycling difficulties. And the *conventional* waste processes for decommissioned turbine blades include landfill and incineration, which are not sustainable solutions. Sustainability does not only include ensuring that a product has as little environmental impact as possible during its use but also that it can be manufactured and disposed of without harming nature; this is a *lifecycle* perspective. Now, think of the huge amount of problematic blade waste that must be recycled in the coming years. Assuming the weight of the blades in Figure 8.2 represents the average weight of existing blades, this could mean roughly one and a half kilograms of composite waste for *every* person living on Earth—until all these blades are recycled. The thought-provoking report by the Norwegian Water Resources and Energy Directorate, titled "Past Management and Future Challenges with Glass Fiber Composites from Wind Turbines in Norway," presents a bleak and broadly applicable assessment of current waste management strategies [15]. It is indeed not easy to recycle something that is glued really hard together.

What about the environmental ethics viewpoint? The NSPE Code of Ethics for Engineers *mentions* the issue of "effective waste management" [10]—but mildly, in the form of encouragement. Besides, there appears to be some sort of cover-up (a form of dishonesty) surrounding the matter, as the general public is not openly informed of the *extent* of the problematic waste issue. The wind power industry has failed in the sustainable waste management of decommissioned turbine blades and fails in light of the virtue of responsibility. On the other hand, "wind power is a renewable and clean source of energy." This is a complicated environmental ethics dilemma.

Fortunately, there is active and promising research and development around new composite materials and advanced recycling techniques for turbine blades [14]. But it will take considerable time to find a sustainable all-around solution and implement it worldwide. Meanwhile, we have to live with the problematic waste issue.

Moreover, could fine-particle dust released from turbine blades due to erosion perhaps emerge as another environmental issue related to wind turbine blades? In line with this concern, a 2024–2027 Danish research project is investigating the environmental effects of *microplastics* released due to the surface erosion of offshore wind turbines (*Source*: wind.dtu.dk/newsarchive/2024/06/project-premise).

Space Debris

Space is endless. Unfortunately, the low and medium Earth orbits, LEO and MEO, as well as the geostationary and equatorial orbit, GEO, are not. At the beginning of March 2024, the satellite tracking website "Orbiting Now" listed 9,494 active satellites in various Earth orbits, and their percentage orbital distribution was [16]:

- 84% in LEO (altitudes 500–1,000 km)
- 3% in MEO (altitudes 5,000–15,000 km)
- 12% in GEO (altitude close to 35,786 km)

These satellites perform a variety of missions, including communications, Earth observation, navigation, space science, and technology development. Some of the critical missions, especially communications and navigation, have a major impact on our daily lives.

Accompanying the active satellites, there is a growing amount of space debris. Based on statistical models produced by the European Space Agency (ESA), it is estimated that there are 36,500 objects larger than 10 cm, 1 million objects between 1 cm and 10 cm, and 130 million objects between 1 mm and 1 cm [17]; and this was in 2022. While traveling much faster than a rifle bullet, those objects can damage anything they hit in orbit. In Figure 8.3, as an example, we can see the Space Shuttle Columbia's flight deck window after its 17-day mission. The cracks radiating from the central point, resembling a spider web, along with several smaller scratches and marks, indicate that the window has suffered from substantial impact and stress. Besides, space debris could even kill an astronaut/cosmonaut/taikonaut

FIGURE 8.3 Documentation of debris impact damage to the Space Shuttle Columbia's overhead flight deck window (NASA ID: sts080-326-010). A point-like damage with spreading cracks caused by a high-velocity impact from space debris is visible in the center of the window. (Photo (19 December 1996) courtesy of NASA.)

on a spacewalk or tear a hole through the International Space Station (average altitude 420 km), for instance. Additionally, the Kessler syndrome lurks there.

The Kessler Syndrome: Once the amount of debris in a particular orbit reaches critical mass, collision cascading begins even if no more objects are launched into the orbit [18].

How big of a problem is this space junk? Space junk is a serious issue and can cause lasting problems for generations to come [17]. So, the cumbersome *waste management* problem is here again, but even more challenging than the turbine blade issue discussed earlier. In fact, the solution to the space debris problem is considerably more complex than the conditions that created it. So far, little progress has been made in solving that expanding problem [19].

But what is the environmental ethics dilemma in this case? Well, more and more satellites are being launched by different countries for various purposes. And we are all increasingly dependent on satellite infrastructure. But as the number of launches continues to increase without effective solutions to the debris problem, we could end up with an environmental disaster in the near-Earth orbits and GEO. This can cause communications and navigation malfunctions, which in turn could lead to severe financial and security issues. Furthermore, reentering space junk threatens people and infrastructure, potentially leaving a wake of destruction on *Earth's surface*. From an ethical point of view, all players in this "game" violate, at least, the virtues of responsibility and respect; there is no adequate safety culture.

Space junk is a truly international *waste management* problem, and therefore, solving it would require active international cooperation and significant co-financing of research and development efforts. And, of course, appropriate internationally binding legislation is needed. Currently, the moral problem is that any space agency can launch material into Earth orbit for their own benefit *without* being held accountable. A diplomatic effort is underway at the UN, led by the United States and the United Kingdom, to establish standards of safe and predictable behavior for all satellite operators. This is an important step in the right direction to get the "tragedy of the commons" under control. The tragedy of the commons refers to a situation where individuals (or organizations, even countries), acting in their own self-interest, deplete or spoil shared resources, leading to negative outcomes for the entire community. Finally, it is good to remember a wise guideline of engineering ethics that "just because we can, doesn't mean we should."

Scenery Damaged by Humans

The *hard* benefits of an air traffic control center considered in this third environmental ethics dilemma are easy to understand and value, but the *soft* costs to nature and people are impossible to quantify. Consequently, all such cost–benefit analyses of the dilemma are necessarily semi-objective—or even subjective. We present this case through a personal vignette, where the existing connection between nature and health (or wellbeing) is recognized from the human rights perspective introduced above.

VIGNETTE It Is My Human Right

A pragmatic engineer, Ovaska is neither an environmentalist nor a human rights advocate. And he committed himself to value-based ethics only in the final years of his 40-year career.

In 2023, he hiked in the Vosges Mountains in France. The highest point of those mountains is the Grand Ballon (1,424 m) in Murbach, and Ovaska wanted to experience it. But that experience was somewhat uncomfortable because there is a radar station on top of the mountain that stands out as a kind of "environmental eyesore" (Figure 8.4). As an electrical engineer and former radio amateur, he understands the purpose and value of radar stations and appreciates

FIGURE 8.4 A bold radar station for air traffic control *versus* the vulnerable beauty of nature. (Photo (1 June 2023) courtesy of Seppo Ovaska.)

the engineering work behind them. However, he got a strong feeling that it is *wrong* to damage such a beautiful scenery. And right next to the radar station is a patriotic World War I monument dedicated to "Les Diables Bleus" of the French Army—so, also the virtue of respect is somewhat in question.

Later, he tried to convince himself that since the Grand Ballon is not in his own backyard, it is really none of his business to criticize the man-made structure on top of it. It surely benefits French aviation. It did, however, affect Ovaska's mental wellbeing during the hiking trip and for some time afterwards.

One morning, when he was already in Germany, he remembered Brei's dissertation, *Our Right to Health and Our Duty to Nature* [12], which he had read a few months earlier. Yeah! Wellbeing—and consequent health (remember the WHO's three-point definition above)—are *his* human rights. He was therefore entitled to criticize the radar station on top of the Grand Ballon; it is clearly an environmental ethics dilemma involving engineering. But is this a selfish point of view? No, because the same applies to anyone who goes hiking, mountain biking, or cross-country skiing in this area. And it probably affects their mental wellbeing to some extent, as well. There must be a good reason why so many people turn to nature when they need some calm and serenity.

In this way, the human rights approach to environmental ethics opened a new door for Ovaska. Maybe for you, too.

From a societal point of view, it was certainly *right* to build the air traffic control center on top of the mountain. There are obvious benefits to having structures and systems in place to make air travel organized and safe. But let us see what the IEEE Code of Ethics (Appendix B) says about this kind of issue: "We, the members of the IEEE… do hereby commit ourselves to the highest ethical and professional conduct and agree: I. To uphold the highest standards of integrity, responsible behavior, and ethical conduct in professional activities. 1. [To] hold paramount the safety, health, and welfare of the public… and to disclose promptly factors that might endanger the public or the environment." It is, indeed, positive to see that this one is in the form of a *commitment*. But did any engineer "disclose promptly factors that might endanger the environment," in the Grand Ballon case? Perhaps, but for nothing.

The Code's connection to environmental ethics issues, such as the Grand Ballon case, is rather weak. On the other hand, why should such issues matter to engineers? A good question, to which we offer the following pragmatic reply: structures like the Grand Ballon radar station may take away from an individual's "wellbeing account" and thus gradually undermine one's personal wellbeing, which has an inherent connection with his *work efficiency* as a practicing engineer (or other professional). Even if one such case has only a marginal negative impact, several similar cases, along with the individual's possible ethical hangovers as well as stressful ad-hoc events and circumstances, will eat away at the balance of his imaginary wellbeing account. And if the balance hits zero, that may lead to health issues. This kind of "wellbeing currency" is analogous to the concept of moral currency discussed in Chapter 4. However, the degree of negative impact can naturally vary among individuals, influenced by personal values and coping mechanisms, as well as professional ethics training (which helps structure ethical issues like that).

An Individual Engineer's Perspective

The environmental ethics dilemmas discussed above differ greatly from other ethical dilemmas faced by practicing engineers if we look at them from the perspective of an individual engineer. Erecting wind turbines, launching spacecraft and satellites, and building radar stations all require a license… or, more accurately, multiple licenses and permits. And these are issued by various agencies (local, state, national, etc.) that base their decisions primarily on applicable acts and laws. So, politicians or legislators ultimately "solve" such ethical dilemmas.

Eventually, engineers use their knowledge and skills to perform *licensed* work in compliance with different standards, professional practices, and their code of conduct. Hence, they are not the "bad actors" behind possible ethical violations during the process of issuing the necessary licenses. Fine? No!—Ultimately, it is every engineer's responsibility to determine whether or not the impact that they have on the world around them is justified by the goods *their works* create. And just about everything that engineers do will have some sort of impact. Thus, they cannot be regarded as outsiders in these cases, as they undeniably bear responsibility for their specific actions.

Those three dilemma cases (wind turbine blades, space debris, and damaged scenery) are certainly interesting and challenging assignments for engineering teams to tackle. And that is what engineers look for and enjoy doing. Most likely, there are no ethical hangovers involved in working with such projects. But it is important to acknowledge that some specific stakeholders and even outsiders might feel rather concerned about these issues.

The pragmatic objective of engineering work is to achieve an economic balance between the amount of pollution or damage *versus* environmental or health aspects. Though, in many environmental matters, engineers are not qualified to make decisions on their own but instead should seek advice from others—such as ecologists, space physicists, and public health experts—who can help analyze and comprehend the potential environmental and health impacts of a technology project. The work that many engineers do can have a tremendous impact on the natural environment and, indirectly, the wellbeing of humans. Failing to take this impact into account is not only a moral failing, it is also a failure of engineering.

SUMMARY

Next, we summarize the sections of this chapter separately and highlight their main lessons.

Disaster

The Chernobyl case study at the beginning of this chapter contains a diverse collection of consequences; but for most of us, one of these can be difficult to relate to the nuclear disaster—namely, the collapse of the Soviet Union, which former President Gorbachev strongly associated with that disaster.

Behind the Chernobyl case was a high-dimensional matrix of actors and numerous violations of ethical principles over several years—up to the terminal action. Thus, the disaster is, indeed, a *meta-case* that does not fit into our straightforward model of unethical behavior (Figure 2.1); but it consists of unethical *sub-cases* that could be explained by the commonsense model.

The core task of the engineers involved was the design of the nuclear reactor and its various operating procedures. But there was no proper safety culture; nor did an appropriate professional code of ethics or an ethical culture exist. On the other hand, the early roots of the NSPE Code of Ethics date back to 1935 [20]. And the current NSPE Code states: "Engineers shall hold paramount the safety, health, and welfare of the public." This *safety* aspect was violated in the design and operation of the Chernobyl reactor [1]. Besides, "Engineers shall perform services only in the areas of their competence." Although the engineers were certainly qualified professionals, their *competence* may have been inadequate when designing such a large-scale, complex *system of systems* that reflects the seamless interplay of engineering, safety, and operational considerations. Thus, the design of the nuclear reactor was flawed, as stated in reference [1].

Basically, nothing could have prevented a disaster like this as long as there was no proper safety culture. Individual engineers had no control over the meta-case; they were victims of management failure. It seems that all was well as long as the reactor worked and contributed to the preservation and growth of public wealth.

It may be hard for you to imagine yourself facing things similar to what happened in Chernobyl. But it is good to remember that *irrational* things do happen every now and then. So, stay awake.

Scandal

Here we move from disaster to scandal. The diesel emissions scandal is a large-scale fraud that was carefully planned, executed, and kept secret—and the fraud remained unidentified for several years. Thus, it fits well with the model in Figure 2.1. From the point of view of consequences, it is a rich case to study; but the individual actor's perspective remains unclear because such information is not (publicly) available. It is, therefore, impossible to complete this unethical puzzle because some of its pieces are missing: the pieces related to the inner workings and motives of the entire chain of actors.

Perhaps the most significant lesson of the Volkswagen case is the observation that unethical cases at the macro-level can, indeed, have *diverse* consequences.—But now, it is time to focus on the positive consequences of the case for society as a whole: "Electric cars are coming. Are you ready?"

Finally, if we dramatize the discovery of this case of deception *tongue in cheek*, we can see an analogy with the well-known story of David versus Goliath from The Bible (I Samuel xvii). Here, Daniel Carder would represent "David" and "Goliath" would be represented by the Volkswagen Group.

No, This Was Not the Last Scandal

It appears that the deceptive actions of large corporations are not over. On 7 January 2021, the Boeing Company was charged with 737 MAX fraud conspiracy, and it agreed to pay over $2.5 billion. "Boeing's employees chose the path of profit over candor by concealing material information from the [Federal Aviation Administration] concerning the operation of its 737 MAX airplane and engaging in an effort to cover up their deception" [21]. And on 7 July 2024, Boeing finally agreed to plead guilty to conspiring to defraud the government.

An in-depth study of this "Boeing case"—an explanation using the tools of behavioral ethics as well as a moral theoretical analysis—would be a recommended student project.

Environment and Ethics

Finally, environmental ethics involving engineers contains a wide variety of difficult dilemmas. Many of these open issues would require substantial engineering effort. Nonetheless, engineers should not be considered as "bad actors" in cases requiring *specific licenses* to carry out an underlying project, such as erecting and commissioning a 100-meter-tall wind turbine next to a busy highway. They just do what some regulatory agency has legally permitted. Therefore, the primary ethical responsibility rests on the shoulders of those who made the issuing of the license *legal*—of course, based on a holistic environmental assessment.

Even though compliance with laws and regulations gives a company its permission to operate in environmentally sensitive projects, it does not guarantee that their operation is ethically sound. Ethical business practices are ultimately in the hands of the organization and its employees. And for that, they need a "moral compass" for intelligent navigation in the vicinity of *right.* This *intelligent* can be defined as an organization's ability to adapt its behavior to meet its business goals in different circumstances.

After the four pragmatic case study chapters, we are ready to present two synthesis-type chapters: Chapter 9 for practicing engineers and Chapter 10 for their organizations.

CLASS EXERCISES

1. A couple of weeks after the Chernobyl accident, ordinary Finns did not know more than that a nuclear accident had occurred, and as a result the radiation level in Finland had increased. To the obvious question: "Why hasn't information been given about Chernobyl and its effects on Finland?" Interior Minister Kaisa Raatikainen (Social Democratic Party) replied succinctly in a live interview: "There is no reason to inform. It only creates fear."

   Analyze her reply using the *Utilitarian* and *Virtue Theoretical* perspectives.

   Class discussion: Compare the students' answers. And, did the Minister's reply reduce or increase fear among citizens? How could she possibly license herself to make such a statement?
2. The two fundamental principles of *doing good* and *doing no harm* are the most immediately relevant to medical practitioners. In their Hippocratic Oath, physicians promise to: "use treatment to help the sick... but never with a view to injury and wrong-doing" [22] (p. 223).

   Nevertheless, a consequence of the Chernobyl disaster was that some physicians advised pregnant women to have abortions due to radiation exposure, even though the levels in question were far below the likely teratogenic effects.

   Did this violate the letter of the Hippocratic Oath above? How could those physicians possibly license this advice to themselves? (Actually, there were many physicians who advised so.) Use one of the AI chatbots as your assistant.
3. In the presentation above, we aimed to convince the reader that the Chernobyl disaster was, in fact, a complex *meta-case.* Could it instead be seen as an *ordinary case* that can be meaningfully explained by the model in Figure 2.1? Justify your answer.
4. Former President Gorbachev made a statement: "Even more than my launch of perestroika, [Chernobyl] was perhaps the real cause of the collapse of the Soviet Union five years later."

   If so, the Chernobyl tragedy brought also a lot of *happiness*: the Cold War ended, the SSRs became independent states, citizens of the former Soviet Bloc got the right to travel abroad, etc.

Could we, therefore, conclude that the Chernobyl case—as a whole—was a positive event instead of a negative one, as is commonly thought?

Class debate: The instructor moderates a prepared debate between two volunteer teams. First, the class is divided into three teams; one of the teams is favoring the positive answer ("as a whole, it was *positive*"), the other is against it ("as a whole, it was *negative*"), and the third team evaluates the arguments presented. What are the objective conclusions of the evaluation team?

5. Daniel Carder and his research team identified and published the diesel emissions discrepancy already in 2014, and EPA became interested in their unexpected observations. This can be seen as a *quiet* beginning of the Volkswagen scandal, which eventually had both negative and positive consequences, as discussed above.

   Analyze the final outcomes of Carder's disclosure from the Utilitarian perspective.
6. Apparently, some manager who was (fairly) high up in Volkswagen's organizational hierarchy gave permission and instructions to proceed with the *defeat device*. This started an extensive unethical process within the company.

   Create a speculative moral self-license for the manager's unethical decision.

   Class discussion: Compare the answers of different students, and assess the strength of their speculated self-licenses.
7. The unethical process mentioned in the previous question consisted of multiple phases and numerous actors: from the initiator of the case to the software engineers who eventually wrote and tested the malicious piece of software.

   Think of the heterogeneous chain of unethical actors; in which link of that chain, if anywhere, were the likely actors who got an ethical hangover from these actions? Justify your answer.
8. Study the Volkswagen Group's code of conduct, Our Code [9], and select the five statements that you consider most important in the light of the previous diesel emissions fraud.

   Class discussion: After completing this engineering ethics course, how long would it take you to internalize *Our Code* if it was the sincere and strong will of your company's top management?
9. Satellite launching and space exploration, in general, are of increasing interest to several countries. And they all understand the space debris problem and its potential consequences.

   Analyze the following ethical issue using one of the moral theories presented in Chapter 3: Does it make ethical sense to continue their space programs without directing major international efforts (engineering and legislation) to the looming space debris disaster?

   Class discussion: Compare the moral theoretical analyses of students and find out what percentage of students think: (a) the current practice is not acceptable (it is *wrong*) or (b) the practice can be continued because everyone else is doing so, too (it is *right*).

10. The text above has the personal vignette "It Is My Human Right." Determine whether it was *right* or *wrong* to build a bold radar station on top of the beautiful mountain (Figure 8.4). Justify your answer.

    Class debate: The instructor moderates a prepared debate between two volunteer teams. First, the class is divided into three teams; one of the teams is favoring the positive answer ("it was *right* to build the radar station"), the other is against it ("it was *wrong* to build it"), and the third team evaluates the profound arguments presented. What are the objective conclusions of the evaluation team?
11. Sustainability of products means that a product has very little environmental impact during use and that it can be manufactured and disposed of without irreparably harming nature. If we consider a battery electric vehicle (EV), its carbon-dioxide emissions during use are practically zero. But mining, refining, processing, manufacturing, and transportation involve significant carbon-dioxide emissions (and other forms of pollution) before we can drive a "clean" EV off a car dealership. In addition, the recycling processes release carbon-dioxide when the EV is decommissioned. And it should be noted that the mining of battery metals and rare Earth metals, commonly used in motor magnets, as well as battery recycling, are all sustainability challenges that are not present in cars with combustion engines.

    Compare the *lifecycle* carbon-dioxide emissions of a midsize electric car and a combustion engine car with average annual kilometers, and find out an estimate of how many years a battery EV should be used before its total carbon-dioxide emissions (*manufacturing + use + disposal*) are lower compared to a conventional car? Use one of the AI chatbots to find your estimate.

    Class discussion: Compare the answers, and try to understand the reasons for their possible differences. Is it ethically sound for the automobile industry and politicians to keep the manufacturing and disposal emissions "under the table?" Could it be seen as a form of dishonesty?
12. Create a mind map of the section "Environmental Ethics."

REFERENCES

1. "Chernobyl accident 1986." World Nuclear Association. Accessed: Jul. 1, 2024. [Online]. Available: https://world-nuclear.org/information-library/safety-and-security/safety-of-plants/chernobyl-accident
2. "Engineering failures: Chernobyl disaster." EIT. Accessed: Jul. 1, 2024. [Online]. Available: https://www.eit.edu.au/engineering-failures-chernobyl-disaster/
3. "Contamination of mushrooms and wild boar with radioactive caesium-137." Gesellschaft für Anlagen- und Reaktorsicherheit (GRS). Accessed: Jul. 5, 2024. [Online]. Available: https://www.grs.de/en/news/contamination-mushrooms-and-wild-boar-radioactive-caesium-137
4. "Why did the Soviet Union collapse?" Britannica. Accessed: Jul. 4, 2024. [Online]. Available: https://www.britannica.com/story/why-did-the-soviet-union-collapse

5. "The research team that caught Volkswagen." Deutsche Welle. Accessed: Jul. 10, 2024. [Online]. Available: https://www.dw.com/en/the-research-team-that-caught-volkswagen/a-18740624
6. J. C. Jung and E. Sharon, "The Volkswagen emissions scandal and its aftermath," *Global Business and Organizational Excellence*, vol. 38, no. 4, pp. 6–15, 2019, doi: 10.1002/joe.21930
7. T. Spapens, "The 'Dieselgate' scandal: A criminological perspective," in *Green Crimes and Dirty Money*, T. Spapens, R. White, D. van Uhm, and W. Huisman, Eds., London, UK: Routledge, 2018, ch. 5, pp. 91–112, doi: 10.4324/9781351245746
8. "Strategy." Volkswagen. Accessed: Jul. 11, 2024. [Online]. Available: https://www.volkswagen-newsroom.com/en/strategy-3912
9. "Our Code." Volkswagen. Accessed: Jul. 12, 2024. [Online]. Available: https://www.volkswagen-group.com/en/integrity-and-compliance-15705
10. "NSPE Code of Ethics for Engineers." NSPE. Accessed: Apr. 17, 2024. [Online]. Available: https://www.nspe.org/sites/default/files/resources/pdfs/Ethics/CodeofEthics/NSPECodeofEthicsforEngineers.pdf
11. C. Palmer, K. McShane, and R. Sandler, "Environmental ethics," *Annual Review of Environment and Resources*, vol. 39, pp. 419–442, Oct. 2014, doi: 10.1146/annurev-environ-121112-094434
12. A. T. Brei, "*Our Right to Health and Our Duty to Nature*," Ph.D. Dissertation. West Lafayette, IN: Department of Philosophy, Purdue University, 2009. Accessed: Apr. 17, 2024. [Online]. Available: https://docs.lib.purdue.edu/dissertations/AAI3379313/
13. "Wind Power Market Size, Share & Trends Analysis Report by Location (Onshore, Offshore), by Application (Utility, Non-utility), by Region (North America, Europe, Asia Pacific, South America, Middle East & Africa), and Segment Forecasts, 2022–2030." Grand View Research. Accessed: Apr. 20, 2024. [Online]. Available: https://www.grandviewresearch.com/industry-analysis/wind-power-industry#
14. J. P. Jensen and K. Skelton, "Wind turbine blade recycling: Experiences, challenges and possibilities in a circular economy," *Renewable and Sustainable Energy Reviews*, vol. 97, pp. 165–176, Dec. 2018, doi: 10.1016/j.rser.2018.08.041
15. S. M. Andreassen, "Past management and future challenges with glass fiber composites from wind turbines in Norway," Norwegian Water Resources and Energy Directorate, Oslo, Norway, NVE Ekstern rapport nr. 22/2023, 2023, 30 p. Accessed: Nov. 28, 2024. [Online]. Available: https://publikasjoner.nve.no/eksternrapport/2023/eksternrapport2023_22.pdf
16. "How many satellites are in space?" NanoAvionics. Accessed: Apr. 22, 2024. [Online]. Available: https://nanoavionics.com/blog/how-many-satellites-are-in-space/
17. "Space explained: How much space junk is there?" Inmarsat. Accessed: Apr. 22, 2024. [Online]. Available: https://www.inmarsat.com/en/insights/corporate/2022/how-much-space-junk-is-there.html
18. "Micrometeoroids and orbital debris (MMOD)." NASA. Accessed: Apr. 27, 2024. [Online]. Available: https://www.nasa.gov/centers-and-facilities/white-sands/micro meteoroids-and-orbital-debris-mmod/
19. J. Tallis, "Remediating space debris: Legal and technical barriers," *Strategic Studies Quarterly*, vol. 9, no. 1, pp. 86–99, 2015. Accessed: Apr. 22, 2024. [Online]. Available: https://www.jstor.org/stable/26270835
20. "History of the Code of Ethics for Engineers." NSPE. Accessed: Jul. 7, 2024. [Online]. Available: https://www.nspe.org/resources/ethics/code-ethics/history-code-ethics-engineers
21. "Boeing charged with 737 MAX fraud conspiracy and agrees to pay over $2.5 billion." U.S. Department of Justice. Accessed: Jul. 9, 2024. [Online]. Available: https://www.justice.gov/opa/pr/boeing-charged-737-max-fraud-conspiracy-and-agrees-pay-over-25-billion
22. S. Bok, *Lying: Moral Choice in Public and Private Life*. New York, NY: Pantheon Books, 1978.

9 Toward Ethical Behavior
Engineers

Learning Objectives

After studying Chapter 9, you will be able to

- Understand that there are no objective reasons not to use value-based ethics as a guideline for practicing engineering
- Realize the potential of team ethical culture in improving individual's local job wellbeing
- Recognize the influence of Generation Z engineers in high-demand fields to steer engineering firms toward ethical business practices
- Recall the various forms and serious consequences of workplace harassment

So far, we have addressed the main title of our book, Ethics for Engineers, from practical and moral theoretical points of view, as well as through numerous micro-level case studies and a few macro-level cases. The following Chapters 9 and 10 have basically absorbed all of that, and it is now time for some synthesis. Chapter 9 discusses such a synthesis from the perspective of engineers, and Chapter 10 examines the organizational perspective in relation to the chapters' core theme, Toward Ethical Behavior. At the end of this chapter, we discuss the tricky issue of workplace harassment, which is a form of unethical behavior that is silenced in many workplaces.

NEW AVENUE FOR EXPERIENCED ENGINEERS

Not Particularly Ethical

At the beginning of this textbook, we aimed to convince students that engineering ethics is, indeed, a useful subject to study. And in subsequent chapters, we highlighted that in addition to being a matter of obligation, ethical behavior is also an opportunity for engineers. Still, many practicing engineers do not behave particularly ethically—and yet they can have more or less "successful" professional careers. But what about their personal wellbeing? Could it be improved by taking moral responsibility for their actions?

We had an opportunity to ask an experienced electrical engineer (who prefers to remain anonymous) why he has *not* adopted value-based ethics as a guideline in practicing his profession. After some hesitation and thought, he gave us the following (somewhat edited) answers, which we believe to be truthful:

1. "Maybe I'm *used to* behaving a bit unethically in certain matters; it's a kind of habit—everyone does it."

 DOI: 10.1201/9781003485520-9

2. "Strictly ethical people are easily *walked over* in career development and promotions—they are naive."
3. "My company's so-called ethics program appears to be *ethics washing* only."
4. "From time to time, I want to *please* my unethically-behaving boss."
5. "Dishonesty is sometimes a good means of *self-defense*."

At first, we were a little confused by his direct answers, but then we went through them one by one and gave him our reasoned views to consider:

1. Virtues are character traits, and they must be acquired through practice. Vices are also character traits and are formed in a similar way. As an experienced engineer, you know well that continuous learning is essential in your profession; likewise, you could also learn virtues such as integrity, fairness, and respect gradually by practicing them.
2. Carrying out your work to a high personal standard is important, and good work and loyalty to shared goals are the real keys to a successful career [1]. Accomplished engineers are not walked over—they are usually not naive. The perception that ethical people are naive may stem from confusing trust with honesty. But it is definitely possible to be completely honest without necessarily trusting others to be the same.
3. Unfortunately, that might be the case. Sissela Bok writes in her book: "But codes of ethics function all too often as shields; their abstraction allows many to adhere to them while continuing their ordinary practices" [2] (p. 246). Nevertheless, you could focus on the ethical work culture of your team—and you can surely influence that yourself.
4. We understand this pragmatic point of view, but there are professional ways to keep your boss happy, such as being a committed and hard-working employee as well as a constructive team player [3]. Make sure your knowledge capital stays up to date. And ask yourself if you are worried about pleasing the right people. Is a career that requires you to routinely go against your values really part of a good life?
5. You may be right, but you should check out Jeremy Bentham's Utilitarian theory and Aristotle's theory of Virtue Ethics (Chapter 3) before deciding when it is appropriate to defend yourself with unethical actions, such as dishonesty. However, this irrational practice must not be overused—be careful. Again, Sissela Bok states in her book: "In principle, then, both deception and violence find a *narrow* justification in self-defense against enemies" [2] (p. 145). (emphasis added)

The anonymous engineer listened to our advice with curiosity and concentration and said finally that he needed to think about what we had to say—and maybe he would give it a try. In fact, ethical behavior is not a "cost" to anyone. It pays for itself in a better quality of life.

Nobody's Benefit

Analyzing the micro-level cases in Chapters 5–7, we found that nearly 70% of the speculated psychological/moral self-licenses were due to the actor's own benefits or benefits to the actor's company (Tables 5.1, 6.1, and 7.1). And about 30% of the self-licenses were classified in the third category, "Other."

From this third category, we selected three cases (Boxes 5.6, 6.4, and 7.5) that highlight the difficulty of seeing how an unethical action could be of any rational benefit to anyone. Such behavior is more likely a strange manifestation of the actor's personality or perhaps even an ego trip. In all these cases, the unethical actor was initially a successful product development engineer who became a manager in his thirties and eventually ended up as vice president. They were apparently considered competent leaders. Let us now present a thought-provoking vignette that contains "ethicalized" versions of the original unethical cases.

VIGNETTE Ethicalized Versions of Three Micro-Level Cases

CONSULTANT'S INVENTION (BOX 5.6)

Recall the case in which Egon ordered Erich to file a patent for an associated user interface that had been designed by a consulting engineer, Gunther. Erich was uncomfortable taking credit for work he had not done, and Egon appeared to be capitalizing on loopholes in the contract with Gunther. We pick up that case near the halfway point and offer an alternate resolution:

> … Erich's boss Egon thought that the novel procedure should definitely be patented, and he asked Erich to start the patent application process immediately. Erich arranged a meeting with Egon, Gunther and the company's patent attorney, where they discussed intellectual property rights (IPR). And after a constructive negotiation, they reached an agreement between Egon's company and Gunther, who was an independent consultant. The IPR matter was handled with the usual professionalism. Gunther was compensated fairly for his invention when he transferred the intellectual property rights to Egon's company. Besides, he suggested that Erich could be the co-inventor, but Erich refused, because he felt that his role in the invention was not that significant. After this positive interaction, their cooperation continued as usual.

RUDE LANGUAGE (BOX 6.4)

Recall the case in which Willi expressed his frustration and disappointment over Kurt's lack of progress by using foul language. We return to the halfway point and continue in a more positive direction:

> … One morning, Willi invited Egon and Kurt to his office—he was finally fed up with the frustrating situation. For more than an hour, they openly discussed the problematic situation, and Kurt explained the reason for his low work energy. He had a newborn baby who kept him awake several hours a night. Willi and Egon understood the challenging situation as they both had two toddlers. They reassured Kurt that things will get better in time.

Being new to object-oriented programming (OOP), Kurt had considerable difficulties with the new way of thinking. Willi suggested an OOP course for Kurt; perhaps first, an introductory course and then, after some practicing, an in-depth course. They discovered that the local college offered such courses in their evening program. In addition, Egon proposed that he and Kurt would practice pair programming during the next few months (pair programming is a technique in which two persons, the driver and the observer, write/review code together). In this way, Kurt would surely become comfortable in the OOP world.

Kurt was happy with these constructive arrangements, and after completing two courses and practicing pair programming for four months, he was considered a professional C++ programmer and proficient user of UML. And after this intensive learning period, he was a grateful and committed employee, too.

THIS IS EXTORTION! (BOX 7.5)

Recall the case in which, during a train trip, Erich offered up an ultimatum, and Walter accused him of extortion. Here, we present a much more constructive alternative:

... Erich mentioned that he would eagerly like to have a multi-year assignment to one of the company's foreign subsidiaries; he thought that would do good for his future career and personal life. Erich shared this idea with Walter and asked his opinion on such a tentative plan. Walter agreed that this kind of foreign experience would certainly be good for Erich and his rising career. And Walter also promised to look around if such an opportunity opens up in one of the company's subsidiaries. Erich said his favorite places would be the United States, Australia, and Canada. Walter grinned and promised to keep that in mind.

After about a year, Walter asked Erich to come to his office and told him enthusiastically that now there would be an interesting opportunity in Australia—for two years. It was a technical advisor position in their Canberra office. Erich was delighted and excited.

...

In all these ethicalized cases, everyone was a winner!

Even though the modified ethical versions are fictional, they are intended to be realistic. In principle, they could have happened if (a) Egon had suppressed his possible personal rancor, (b) Willi his need to emphasize his ego, and (c) Walter his sensitivity to the outspokenness of his subordinate. Thus, there was no rational reason for their unethical behavior. And in the future, Egon, Willi, and Walter could safely choose the ethical avenue without losing anything. However, the obvious connection of these unethical actions to the actor's ego may cause trouble because the ego is closely tied to their personality.

The earlier in their career a professional commits to behaving and acting ethically, the easier it is to adopt such behavior as a true personal quality—as "old dogs don't learn new tricks." Moreover, this kind of manager/employee commitment is also a cornerstone of an ethical company culture, as stated in Chapter 4.

REDUCING UNETHICAL BEHAVIOR

In this section, we look at the possibilities to reduce unethical behavior in cases where engineers are involved. Both micro-level actions at the team level and macro-level actions beyond one's own team are considered. In addition, a hopeful hypothesis is presented, and it is proposed as a future research topic for behavioral ethics scholars.

TEAM ETHICAL CULTURE

Engineering, like many professions, is a team sport; practically no engineering problem can be solved by one engineer alone. Many engineering organizations are divided into sections or departments and further into groups or teams. Most engineering teams have no more than 10–20 members, and larger units, or formations of several teams, are called sections or departments that consist of several tens of employees—even more. While an individual engineer may not have much influence on the ethical culture of the entire organization, she can have a positive impact on the ethical culture at the *team* level.

Let us take an example related to wellbeing at work: if any team member has a *negative* behavior pattern, it easily weakens the working atmosphere of the whole team. Therefore, if virtues such as fairness, honesty, integrity, respect, or responsibility are regularly absent, the resulting unethical behavior affects the wellbeing of all team members—and thus the overall work effectiveness of the team.

Similarly, if the behavior of the supervisor or an accomplished (as measured by expertise and skills) team member is consistently ethical, she creates a *positive zone* around herself. And since teams are small, the interaction between members is active and regular. Hence, the ethical behavior of this respected team member is transmitted to her closest team colleagues... and further to their closest colleagues, and so on. Eventually, there is a notable likelihood that the entire team will begin to adhere to similar ethical values. Such self-emerging team ethics is illustrated in Figure 9.1, where the ethical values of one team member gradually spread to all members. As a result, the behavioral culture at the team level becomes ethical, which leads to improved *local* workplace wellbeing—even if the engineering organization or company the team belongs to does not have an active ethics program or particularly ethical business practices. For this reason, the ethical behavior of a single team member can turn out to be a local opportunity for positive change. And if your company does not have a code of conduct, you can adopt the IEEE Code of Ethics (see Appendix B) as a professional guide for your team.

However anecdotal this may sound, it is supported by insights from psychology, sociology, philosophy, and the various profession-focused applications of those disciplines. Just reflect on your own experiences of group dynamics, and you will surely see that attitudes and actions are contagious. Doubtless, you will also see that positive attitudes and good actions are constructive, while negative attitudes and bad actions are destructive. Further—and perhaps of more interest to the empirically-minded—support comes from Cabana and Kaptein, who discuss similar issues in

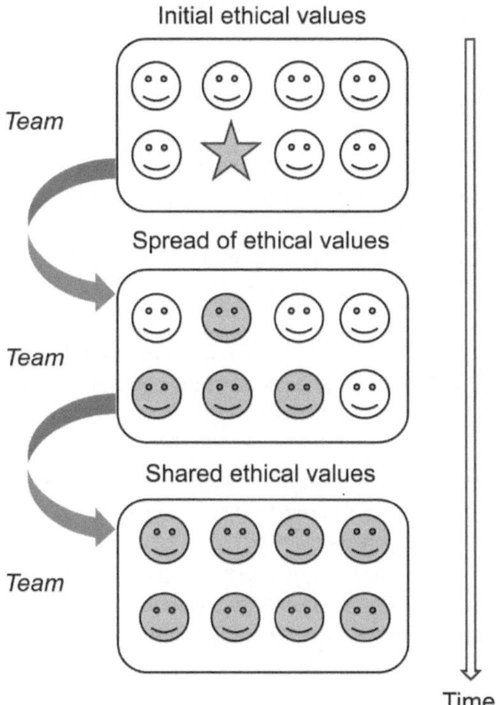

FIGURE 9.1 Self-emerging team ethics; the gray star and gray faces represent team members whose actions are grounded on ethical values, and white faces represent members who act impulsively and arbitrarily.

their article on Team Ethical Cultures within an Organization [4]. They wanted to find out if different teams in an organization can have different ethical cultures, and for this purpose, they surveyed 180 teams in one large British organization. And they found out that there were significant differences between the ethical cultures of individual teams. In addition, the ethical aspects of team culture were transmitted from member to member through socialization processes, such as *modeling*, *observation*, and *interaction*. As a result, the more closely team members interact, the more they tend to develop shared values and behaviors. It is in this way that the ethical behavior of the single team member spreads to the entire team, as shown in Figure 9.1 (depicted in three temporal illustrations of the team's ethical state). It was already mentioned in Chapter 4 that "actions of other people can influence one's own actions in the ethics domain."

To conclude, even if your company is not committed to ethical business practices, it is still important that *you* act ethically within your team because your ethical values can be transmitted to other team members. The result could very well be that the whole team could develop an ethical work culture, which makes everyone involved better off—including *you*. Trusted supervisors or team members are often role models for their colleagues. Furthermore, such ethical role modeling is another cornerstone of an ethical company culture (Chapter 4).

Beyond One's Own Team

When engineers have learned and become accustomed to behaving ethically in their team environment, they have considerable internal inertia to push past the temptation to ignore ethical issues or to choose immoral actions, even beyond their team context. Based on this insight, we came up with a hypothesis that recommends *hope* concerning companies with a low ethical culture.

> **Hypothesis:** As engineers learn to manage their small ethical challenges, the number of bigger unethical cases they become involved in gradually decreases.

This basically means that once engineers are used to behaving ethically in their team environment (and with small, day-to-day issues in general), this kind of behavior will become *their way* of acting and will set them up well for dealing with bigger issues beyond their team. Of course, big unethical cases are often complex and involve many individuals. Assuming a number of these individuals have already learned to manage their small ethical challenges, then the number of big unethical cases involving engineers could gradually decrease over the long term because unethical behavior is *not* their way (because it goes against their values). Nonetheless, a company's unhealthy ethical culture makes the matter a bit complicated because an employee who fails to abide by her employer's unethical standards might be denied career advancement opportunities and may even lose her job. This is a puzzling value-dilemma, to be sure. Then again, searching for another employer can be an option worth considering. But since we can neither anticipate every possible situation nor provide a full analysis of the values involved in these kinds of dilemmas, we leave the above hypothesis as an open research question for behavioral ethics researchers. (And we look forward to the results with optimism.) Additionally, all practitioners together could make this hypothesis a reality now—if they only wanted to.

GENERATION Z BOOSTS ENGINEERING ETHICS?

Technology Development and High-Demand Jobs

Speaking generally, Generation Z (*Gen Z*, for short) engineers and engineering students (born 1997–2012) place significant value on building relationships and expanding their knowledge base, both of which are important to their personal and professional growth, especially for early-career employees. They also value things like freedom, flexibility, equality, tolerance, environmental issues, minimal bureaucracy, telecommuting, and "feeling good" in general. And money may not be their crucial motivator—especially for those engineers who are aware that they are in *high-demand* in the job market; but they do appreciate the intangible benefits, like an ethical work environment.

Microprocessor (1971)

Personal computer (1977)

World Wide Web (1990)

Digital cellular technology (1991)

Google search engine (1998)

Touchscreen smartphone (2007)

AI chatbot (2022)

Gen X 1965–	Gen Y 1981–	**Gen Z** 1997–	Gen α 2013–28

FIGURE 9.2 Generations around Gen Z and major ICT (information and communications technology) advances from 1971 to 2022. Note: Gen Y is widely known as Millennials.

Figure 9.2 shows the 16-year generation time spans from Gen X to Gen Alpha (α), as well as parallel steps in technological development. Note that some of the years of these technological steps are not unambiguous and should therefore be considered approximate. In fact, technological development is currently faster than the ability of an individual's *responsibility* to adapt to rising new possibilities. This poses novel ethical challenges to society as a whole since we all have free will, which—in engineering terms—requires the adaptation of a stabilizing control loop, which takes some time.

Examples of currently "hot" engineering fields with significant growth and a shortage of skilled workforce include the following [5]:

- Artificial intelligence
- Biomaterials
- Computer cartography
- Cybersecurity
- Electric vehicles (land, sea, and aerospace)
- Embedded systems in healthcare
- Nanotechnology
- Quantum computing
- Renewable energy systems
- Wearable sensors

Such a list is inherently dynamic and may even change suddenly over time. While employers are already accustomed to paying high salaries to attract and retain top talent in "hot" engineering fields, the playing field is somewhat different with Gen Z professionals, who may be more interested in things like an *ethical organizational culture* (and, relatedly, job fulfillment, work/life balance, and respect) than cold, hard cash.

Moreover, it should be mentioned that many of the high-demand fields are theoretically demanding and would therefore benefit from engineers with graduate degrees. These fields include artificial intelligence, nanotechnology, and quantum computing, as examples from electrical and computer engineering.

A New Kind of Value Base

Gen Z is characterized in a survey report by McKinsey & Company as "[a] hypercognitive generation very comfortable with collecting and cross-referencing many sources of information and with integrating virtual and offline experiences" [6]. And regarding their *personal values*, the same report expresses: "Businesses must rethink how they deliver value to the consumer, rebalance scale and mass production against personalization, and—more than ever—practice what they preach when they address marketing issues and work ethics." Hence, the majority of Gen Z people tend to buy products from companies they consider to be *ethical*. Social media and AI chatbots help them easily find violations of corporate ethical behavior, to which they can react immediately. Based on their ethical consumer attitude, it can be anticipated that Gen Z engineers would also value employers with an *ethical work culture*. Not only would they value such an employer, but they would likely select jobs and make demands based on this criterion. Previous generations of engineers rarely made this a priority. Of course, Gen Z is not homogenous; but what we call here "Gen Z engineers" represent the typical engineers of that generation *in their core behavior*. This behavior displays a concern for justice, an appreciation of individuality, a reliance on dialogue, and a pragmatism rooted in truth [6].

Even though Generation Z engineers appear to be a sort of *ethical opportunity* for the engineering profession, they create challenges for the development of appropriate management practices. New leadership styles and strategies are needed to maximize the value of Gen Z engineers to companies and to keep them engaged with the company's business goals. There will be a challenging generational gap in engineering organizations in the near future: Generation X supervisors (born before 1981) versus Gen Z subordinates, who may have very different values. Of course, there may be risks of stereotyping this particular demographic. In time, we might be able to remove the question mark from the header of this section.

WORKPLACE HARASSMENT

Workplace harassment belongs to the broad *Occupational Safety and Health* category [7]. There have long been *standards* and *regulations* that protect employees in their work environment. A typical example of these is the maximum permissible amount of certain toxic chemicals in laboratory air. The concentrations of such chemicals are measured regularly, if not continuously. This sort of monitoring has been conducted for years. On the other hand, it is much more difficult to measure the existence and severity of workplace harassment; there are no specific sensors for that purpose. This is one reason why workplace harassment may go unrecognized and may not be taken as seriously as, for example, exposure to toxic chemicals.

Future vision: We do have a bright vision of the harassment sensor of the future. One day, smartwatches combined with body and environment sensors, as well as generative AI connected to social media platforms, could provide us with *virtual* harassment sensors, which would be monitored at workplaces. Talk about an ethical frontier! With advanced technology like this (using data analytics), moral character becomes more important than ever.

A Tricky Form of Unethical Behavior

But what is workplace harassment all about? Actually, it is not a monolithic issue but can take many forms, including those listed below:

- Bullying of various sorts
- Discrimination (such as age, gender, race, or sexual orientation)
- Humiliation (traditionally or in cyberspace)
- Isolation (from activities, conversations, and interactions)
- Mockery (belittling or ridiculing)
- Sexual harassment in all forms
- Subjugation (intimidation or manipulation)

And in these cases, the behavior involved can relate to such virtues as fairness, loyalty, respect, tact, tolerance, and truthfulness. Fundamentally, harassment amounts to a disregard for human value. Furthermore, new means of harassment emerge every now and then; social media is a versatile platform for workplace harassment. To give an example of traditional workplace harassment, we next present a quasi-factual vignette from the working life of an engineer who worked as a robotics specialist. This harassment was perpetrated by another engineer, and although it might not be considered extreme, it affected the harassed engineer's wellbeing at work and thus his overall work performance. For these and other reasons, it is a matter of serious concern.

VIGNETTE You're Doin' Our Job

Kurt worked as a systems engineer in a European corporation that had acquired a medium-sized company in Alabama, USA. Their main business field was industrial robotics, and Kurt was transferred from the parent company to this new subsidiary for one year. During that period, he was tasked with developing control algorithms for an ultra-light hand of a special assembly robot. Kurt moved to Alabama full of enthusiasm with his young wife.

Initially, Kurt worked in the company's robotics laboratory, where he developed and evaluated advanced control algorithms for the robotic hand. He worked mostly alone because nobody in the subsidiary had experience with adaptive neural-network controllers. His workdays were intense and long.

But there was Walter, a nasty local engineer, who did not like Kurt. Although they had nothing professional to do with each other, Walter regularly spent his coffee breaks in Kurt's lab, verbally harassing Kurt. Walter thought they did not need help from any Europeans, and he made this viewpoint very evident. He was certain that his company could get along just fine without Kurt. Walter took Kurt's presence to suggest that Europeans were smarter than people from Alabama, an idea he very strongly rejected. Moreover, Walter criticized Kurt for spending so much time at work, calling him a zombie. He even suggested that Kurt's long hours at work meant he was not performing his husbandly duties and that Kurt's wife was being neglected. Occasionally, Walter would slip a travel brochure under the windshield wiper blade of Kurt's red Camaro, advertising cheap plane tickets to Europe. Subtlety was not one of Walter's strong suits.

This scene was replayed over and over again during Kurt's tenure. Fortunately for Kurt, Walter was a field engineer and spent a few days each week out of town. Kurt surely appreciated those breaks. However, he felt moderate work stress about regular verbal harassment and often got migraine headaches. He did not understand why Walter had to act like that.

So, why was Walter wasting his coffee breaks harassing Kurt, whom he did not even know? And more generally, why would anyone want to harass another person? In Walter's case, perhaps it resulted from personal or professional insecurity. Perhaps he only felt good about himself by making others feel bad about themselves. Or perhaps Walter became addicted to harassing others in a similar way that an alcoholic is addicted to booze. Typical reasons why someone engages in workplace harassment are given below [8]:

- Envy (to harm)
- Fears (to assert dominance)
- Intolerance (to discriminate)
- Rancor (to revenge)
- Self-esteem problems (to highlight one's own importance)
- Rivalry (to outperform)

When Walter harassed Kurt, he failed to exhibit at least the following virtues: fairness, respect, tact, and tolerance. And Walter's vicious behavior caused Kurt to feel work stress and suffer from headaches. Hence, Walter's recurring verbal attacks were far from harmless joking or "boys' talk."

It would be interesting to know if there is any connection between workplace harassment and previous harassment behavior at school (K–12). At school, an underlying reason behind harassment is often the desire to be accepted and to belong. This may be the ultimate goal of both the harasser and the harassed, in fact. And if such a connection existed, it would mean that the harasser's unethical behavior in the workplace may have relatively long roots. This would be a suggested research topic for behavioral psychologists. Our suspicion is that some habits of character are established at quite a young age and that some people never mature beyond their childish ways.

A Gloomy Set of Consequences

Heinz Leymann stated in a Swedish survey report that "one in seven adult suicides are as a result of workplace bullying" [9]. This is a dramatic claim, but it should not be generalized or specifically directed at engineers, as the study is based on interviews in Sweden only. Still, it is a striking signal to everyone. Whenever national reports of workplace harassment are studied, it must be noted that there are considerable differences in international work cultures. Rayner and Hoel discuss the issue of workplace bullying in their pioneering review of Literature Relating to Workplace Bullying with over 60 references [10].

Because harassment comes in many forms and degrees of severity, this type of unethical behavior has several possible consequences. Figure 9.3 shows some of the

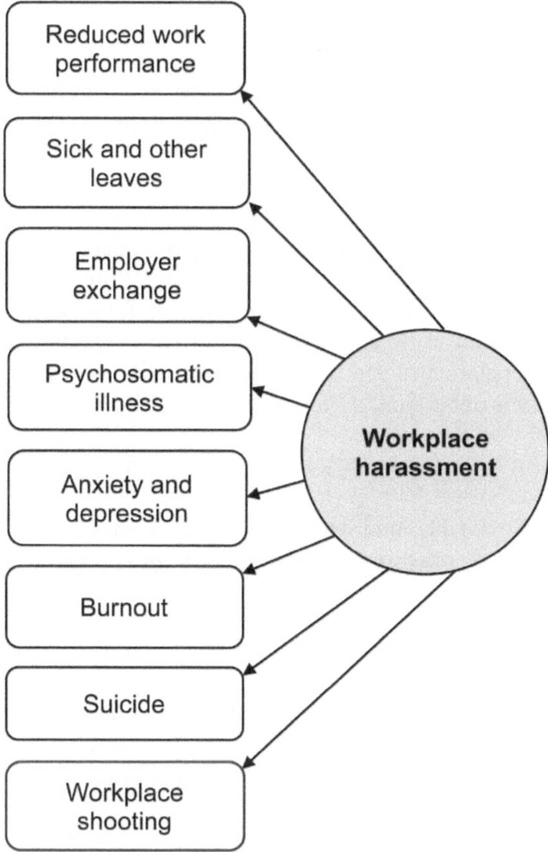

FIGURE 9.3 Possible consequences of workplace harassment for a harassed employee.

consequences for harassed individuals, which also affect the local organization—a team or a department. From this eye-opening diagram, we can see that the dynamics of the consequences are wide, ranging from *minor* to *dramatic*. And if the surrounding organization and management tolerate the harassment silently, then the organization itself must be sick.

Based on our observations in a few engineering organizations and conversations with a handful of harassed engineers, it seems that there are usually only one or sometimes two harassers—rarely more. Nonetheless, the situation may be different among blue-collar employees in manufacturing and maintenance jobs. Since these particular engineering companies rarely have an ethics program in place, management often takes no action to prevent harassment. Very likely, they do not even know that such a thing is happening among their subordinates. And, of course, whistleblowers ("informants") rarely step forward to inform management about the harassment taking place in their organization. The unethical behavior just continues, and the consequences worsen.

Fortunately, a company-wide ethics program also gives hope and possible relief to troubled engineers who may have been victims of persistent unethical treatment. And we can all play a part in relieving the problem of harassment by making sure we do not harass anyone ourselves. Besides, it is definitely recommended to *blow the whistle* even without an existing ethics program—be brave and help your colleague! In the spirit of President John F. Kennedy's legendary words: ask not what your company can do to end workplace harassment—ask what you can do to end it.

By the way, the well-known *MeToo* movement focused on sexual harassment in workplaces [11]. The profession of engineering would benefit greatly from similar critical attention on *all* forms of workplace harassment. Eliminating the hazing, tribalism, sexism, and other forms of harassment typically found in engineering would go a long way toward ensuring a safe and supportive work environment. It would also enhance the reputation of the profession itself, which is the responsibility of every engineer. Think about that.

SUMMARY

Toward ethical behavior has multiple layers when viewed at the individual level, especially in engineering organizations *without* an ethics program. This includes the four distinct layers discussed in this chapter:

1. Experienced engineers who do not have a particularly ethical code of conduct.
2. Team members who can influence the ethical culture of their own team.
3. Generation Z engineers who hold new kinds of core values.
4. Engineers who are involved in workplace harassment.

Each of these layers has its unique characteristics, as well as opportunities and constraints. We hope that their diverse themes provide an inspiring basis for class discussions.

In the first layer, experienced engineers who work "ordinarily" without ethics explicitly in mind have the potential to transform into values-based ethics practitioners. They may not have thought about it at all, or they may have a false preconception that they would *lose* something by holding themselves to a higher ethical standard. A pragmatic course in engineering ethics combined with open peer discussions could remove the barrier of prejudice.

The second layer gives hope to team members who would appreciate a more ethical company culture but cannot influence it. While one individual is unlikely to change the culture of an entire company (if he/she is not the CEO!), they could focus instead on their team's ethical culture. They can become pathfinders as the ethical aspects of the team culture are passed on from member to member through socialization processes. In this way, consistent ethical behavior is gradually transmitted to other team members, and the end result can be an ethical work culture at the *local* level.

Generation Z engineers form the third layer. They are a digitally-native generation, sometimes called influencers. In general, the value base of Gen Z engineers seems to be less materialistic and more humane than that of their predecessors. This prompts them to pay close attention to the company's ethical culture before signing on the dotted line for their first full-time engineering job; money may not (and should not) be their primary driver. An executive of a large company acknowledged to us signs of such emerging behavior among recently hired engineers and business graduates. And particularly in the *high-demand* engineering fields, this can create pressure for companies that do not already have an ethics program.

Finally, the fourth layer, related to workplace harassment, is very difficult to deal with on an individual level. It is (almost) necessary to rely on the professional help and authority of management. Because their mission is to maximize the performance of engineers and the efficiency of the organization, they must also be alert to potential harassment within their department. But unfortunately, that is easier said than done to get positive results. Therefore, we reluctantly pass this tricky topic to the next chapter, which examines the same theme as this one but from the perspective of *organizations*. Could Chapter 10 also provide a practical solution that would work for workplace harassment?

CLASS EXERCISES

1. We asked an experienced engineer, *why has not he adopted value-based ethics as a guideline in practicing his profession*? And he listed five points for his answer at the beginning of this chapter. Use one of the AI chatbots to find an answer to the same question. What are the main similarities and differences between the anonymous engineer's response and the response from your chosen AI chatbot?
2. Egon (Box 5.6), Willi (Box 6.4), and Walter (Box 7.5) behaved unethically in their respective case stories. How could their future behavior in similar situations be transformed to be *ethical* (see the ethicalized stories above)? Or is this a hopeless objective?

   Class discussion: Compare the answers of different students and create a common answer for the whole class.

3. In the section of Team Ethical Culture, we proposed that ethical behavior of an *accomplished* team member may be transmitted to team colleagues. Why should this "seed member" be particularly accomplished? Why would *any* team member not have the same impact?
4. In the text above, we presented a hypothesis that creates hope: *as engineers first learn to manage their small ethical challenges, the number of bigger unethical cases they become involved in gradually decreases.* Is this hypothesis true or false?

   Class debate: The instructor moderates a prepared debate between two teams. First, the class is divided into *three* teams; one of the teams supports the positive answer ("the hypothesis is true"), the other opposes it ("it is false"), and the third team evaluates the arguments presented. What are the objective conclusions of the evaluation team?
5. First, study McKinsey's Generation Z report [6]. Since you may belong to that generation, would you be willing to sign the report? If not, what are the most contradictory results from their survey?
6. The section titled *Generation Z Boosts Engineering Ethics?* suggests that Gen Z engineers could be an *ethical opportunity* for the engineering profession. Could that really be the case? Why/why not?

   Class discussion: Would the early-career engineers of Gen Z be more ethical than their predecessors?
7. What could have been the reasons why Walter harassed Kurt in the vignette above? Create for him a speculative moral self-license that would allow such unethical behavior. For your information, Walter was the only one who harassed Kurt; everyone else was really friendly.
8. Why do only certain people (a vast minority) harass other people in engineering workplaces, i.e., how is a harasser "born?" Use an AI chatbot as your assistant to create a thoughtful answer.
9. Create a mind map of this chapter.

REFERENCES

1. J. Treichler, "A team sport," *IEEE Potentials*, vol. 37, no. 3, pp. 43–45, 2018, doi: 10.1109/MPOT.2018.2794660
2. S. Bok, *Lying: Moral Choice in Public and Private Life*. New York, NY: Pantheon Books, 1978.
3. S. J. Ovaska, "Managing your career in a dynamic environment," *IEEE Potentials*, vol. 37, no. 3, pp. 24–26, 2018, doi: 10.1109/MPOT.2017.2764512
4. G. C. Cabana and M. Kaptein, "Team ethical cultures within an organization: A differentiation perspective on their existence and relevance," *Journal of Business Ethics*, vol. 170, pp. 761–780, 2021, doi: 10.1007/s10551-019-04376-5
5. S. J. Ovaska, "If I were a student again: My next choice," *IEEE Potentials*, vol. 37, no. 3, pp. 41–42, 2018, doi: 10.1109/MPOT.2017.2734941
6. "True Gen": *Generation Z and Its Implications for Companies*. Sao Paolo, Brazil: McKinsey & Company. Accessed: Mar. 8, 2024. [Online]. Available: https://www.mckinsey.com/industries/consumer-packaged-goods/our-insights/true-gen-generation-z-and-its-implications-for-companies#/

7. M. A. Friend and J. P. Kohn, *Fundamentals of Occupational Safety and Health*, 8th Edition. Lanham, MD: Bernan Press, 2023.
8. S. Gordon, *Why People are Bullied at Work*. New York, NY: Verywell Mind, 2023. [Online]. Available: https://www.verywellmind.com/reasons-why-workplace-bullies-target-people-460783
9. H. Leymann, *Psykiatriska problem vid vuxenmobbning: en rikstäckande undersökning med 2438 intervjuer ('Psychiatric problems in adult bullying: A nationwide survey with 2438 interviews')*, Arbetarskyddstyrelsen. Stockholm, Sweden: Delrapport, 3, 1992, 23 p. [In Swedish].
10. C. Rayner and H. Hoel, "A summary review of literature relating to workplace bullying," *Journal of Community & Applied Social Psychology*, vol. 7, pp. 181–191, 1997, doi:10.1002/(SICI)1099-1298(199706)7:3<181::AID-CASP416>3.0.CO;2-Y
11. M. Rodino-Colocino, "Me too, #MeToo: Countering cruelty with empathy," *Communication and Critical/Cultural Studies*, vol. 15, no. 1, pp. 96–100, 2018, doi: 10.1080/14791420.2018.1435083

10 Toward Ethical Behavior
Organizations

Learning Objectives

After studying Chapter 10, you will be able to

- Understand the content and objectives of corporate ethics programs
- See the potential and obstacles to a uniform business ethics standard and certification
- Appreciate the presence and contributions of ethics ambassadors
- Latch on to the fact that corporate ethics programs are free in the long run
- Realize the importance of consistent ethical behavior when an organization faces a crisis

The mission of this final chapter is to guide engineering organizations toward (more) ethical behavior that would also enhance the *moral wellbeing* of individual engineers and engineering managers. Basically, same approaches and methods apply to virtually all types of organizations. To accomplish our mission, we first discuss a top-down strategy for instilling ethical business practices in organizations. We then introduce two service roles that organizational members can volunteer to perform, namely the ad-hoc whistleblower role and the persistent ethics ambassador role. After this, we consider the essential question, could business ethics programs be free for companies—or even profitable—in the long run? And finally, workplace ethics is discussed in crisis situations, especially in those where companies are forced to downsize their operations and lay off employees to cut expenses due to challenging economic conditions.

CORE VALUES TO BUILD ON: TOP-DOWN IMPLEMENTATION

Top-down implementation of ethical business practices appears to be a natural choice according to the sections "Unethical Organizational Culture" (Chapter 2) and "Unethical Behavior within Organizations" (Chapter 4). In fact, it is hard to imagine any other implementation strategy that would work, except maybe in small *agile* organizations. In a special environment like that, the structure, practices, and capabilities of the organization are planned in such a way that employees can react rapidly (or adapt) to changing operating conditions. In software development, for instance, agile processes harness change for the customer's competitive advantage. Therefore, the implementation strategy of business ethics in an agile organization could perhaps follow an alternative *case-by-case* approach, as well. This might be a

DOI: 10.1201/9781003485520-10

fresh research topic for business ethics scholars. Either way, the end result is a sort of company-wide ethics program.

Nordic Business Ethics Survey

What is the state of business ethics in companies today? Well, the answer depends on who you ask, of course. Cultural differences, economic conditions, various traditions, and types of business—among other things—affect the state of business ethics in different countries and within specific companies. But as we wanted to get some answer to this tempting question, we turned to the *Nordic Business Ethics Survey* [1], which represents fairly uniform Northern European countries with a similar culture, economic situations, and traditions. That study is a kind of buttress for this chapter, as it provides insight into how employees perceive the ethical state of their company. Nevertheless, the results shown should not be generalized beyond the Nordic region without careful consideration.

For a brief introduction, this "Nordic" includes Denmark, Finland, Norway, and Sweden. The total population was 27.5 million in 2023, and their Gross Domestic Product per capita averaged about $66,500, which is 18.6% below the United States (*Source*: data.worldbank.org). Furthermore, these Nordic countries have a well-known tradition of ethical behavior, where trust in public institutions is high, corruption in the public sector is low, and the news media is free.

TNS Kantar, a global market research and information group, collected the data in February 2020, just before the outbreak of the COVID-19 pandemic. A total of 4,211 responses were collected (1,007–1,116 from each country), of which 50.5% were from men and 49.5% from women, with various job descriptions, and to a large extent from companies with more than 250 employees [1].

We present some subjectivity-selected, "interesting" results of the survey in Table 10.1. But instead of giving the mean percentages of responses from different countries—as they did in the survey report—we give the medians, as a median value is robust to outliers. This four-point median operator ignores the minimum and maximum values and returns the average of the two values between these extremes. In this way, we aim to reduce the effect of possible country-specific bias in responses; its effect on these particular results is at most a few percentage points. For example, response 2 contains a significant outlier: Denmark 61%, *Finland* 37%, Norway 63%, and Sweden 62%. Consequently, the median is 61.5%, and the mean is 55.75%—their difference is 5.75 percentage points. And Finland's strikingly low percentage might be an indication of a less mature ethical business culture compared to other Nordic countries, but this cannot be confirmed only on the basis of this single observation.

The selected responses 1–5 represent the *ethical foundation* in companies, while 6–12 are explicitly related to *unethical behavior*. Rather than discuss these responses here, we refer to each of them in later sections of this chapter.

But in brief, how is business ethics doing in the Nordic countries? After studying the survey report and combining its results and conclusions with our personal

TABLE 10.1
Selected Responses from the Nordic Business Ethics Survey 2020 [1], along with the Rounded Median Percentages of Corresponding Responses from Different Participating Countries

Response	Median %
1. Doing ethical business is already financially rewarding	48
2. My organization has a code of conduct or code of ethics	62
3. At my workplace, training on ethical workplace behavior is organized regularly	29
4. A web-based whistleblowing tool is available, which enables anonymous reporting	17
5. I strongly agree that my manager (line manager/direct supervisor) promotes integrity and doing the right thing in his/her talks and presentations	28
6. My manager was involved in activities that are against the company's code of ethics	49
7. I have found myself in a situation at work where I had to compromise my personal ethical standards	31
8. Following my manager's instructions was the main reason for compromising my personal ethical standards	21
9. I have found myself in a situation at work where I had to compromise the ethical standards of my organization	15
10. Following my manager's instructions was the main reason for compromising the ethical standards of my organization	22
11. I have observed discrimination or bullying within my company in the past 12 months	34
12. I did not intervene when I witnessed unethical or illegal conduct	64

Source: Reprinted with the Permission of the Nordic Business Ethics Initiative.

experiences and confidential discussions with some engineering practitioners, a semi-objective grade could be C+ ("moderate"). Nonetheless, it should be noted that there can be significant variation between companies around this generalization. Anyway, there is still much room for improvement.

Business Ethics Programs

Content and Objectives

To begin with, an ethics program is *a formal system of organizational control designed to prevent unethical behavior.* An insightful study by Muel Kaptein showed that "unethical behavior occurs less frequently in organizations that have an ethics program than in organizations that do not have an ethics program" [2]. This is a hopeful guidepost for companies struggling with unethical business practices; indeed, there exists a trail to ethical behavior.

Furthermore, Kaptein's research revealed that the *best* order to implement the components of an ethics program is as follows:

1. A code of ethics
2. Ethics training and communication

3. Accountability policies
4. Monitoring and auditing
5. Investigation and correction policies
6. An ethics office(r)
7. Ethics report line
8. Incentive policies

The first five components are *directly* related to less unethical behavior, while the last three are *indirectly* related to less unethical behavior. This questionnaire research was based on a comprehensive dataset of 5,065 respondents from different organizations in the United States. For more information, the reader is directed to the original article by Kaptein [2].

It should be noted that, due to budget and time constraints, an ethics program is often implemented in a few steps instead of all the way; or perhaps only the most significant subset of the components listed above is implemented at all. And this "significance" is an organization-specific quality. However, the first implemented subset contains usually the *code of ethics*, an *ethics office(r)*, and *ethics training and communication*.

Kaptein reckoned that "the frequency of unethical behavior in organizations is better explained by the context of the organization than by the attitude of individual employees." And he defined the principal functions of ethics programs, which aim to stimulate ethical behavior [2]:

- Offer employees clarity about (un)ethical behavior
- Demonstrate management's role model behavior
- Provide employees with the resources they need to behave ethically
- Promote employee engagement in ethical behavior
- Increase transparency around employees' (un)ethical behavior
- Create openness in the discussion of ethical issues
- Reinforce the ethical behavior of employees

As a comment related to the *management's role model behavior*, only 28% of respondents of the Nordic Business Ethics Survey strongly agreed that their managers promote integrity and doing the right thing in their day-to-day interactions [1] (see Table 10.1). On the other hand, 49% of respondents stated that their managers were involved in activities that were *against* the company's code of ethics. Thus, managers lead with both a stimulating attitude and a negative example—but hopefully not the same individuals, as that would send an absurd signal to their subordinates.

Kaptein's article outlines a general-purpose ethics program and a sequence of its implementation. However, it is common for companies to *tailor* their own ethics programs to the specific needs arising from their operating environment. Hence, we invited Pontus Selderman, Senior Vice President, Ethics and Compliance, from Stora Enso (www.storaenso.com/en) to author Appendix A with the title "Ethics and Compliance Program in a Major International Company." His contribution complements Kaptein's point of view in reference [2] and offers a genuine company viewpoint to consider.

Implementation

How should the implementation of an ethics program start? Usually, the first component that is implemented is the *code of ethics*. Creating such a strategic document is a substantial effort, but first, the company must establish its values, as they form the basis of the code of ethics or the broader code of conduct. As an example, the *group essentials* are the foundation of the Volkswagen Group's values (see Table 10.2) and are thus the cornerstones of *Our Code*, their in-depth code of conduct [3].

We can further distill these group essentials to plain values, for instance, as follows: (I) responsibility, (II) honesty, (III) creativity, (IV) respect, (V) assertiveness, (VI) cooperation, and (VII) trustworthiness. These values form the moral basis of their code. In the Nordic Business Ethics Survey, 62% of respondents confirmed that their organization has a code of conduct/ethics—which is still a fairly modest proportion; but maybe it can be explained by the varying size of the organizations studied.

Further, in the Nordic Business Ethics Survey, only 15% of the respondents confirmed that they had been in a situation where they had to compromise their organization's ethical standards—that's good. But surprisingly, 22% of them stated that their manager's instructions were the main reason for acting so—that's bad. The corresponding figures for violating personal ethical standards at work were 31% and 21%, respectively. It would be interesting to know what proportion of that notable 31% resulted in an ethical hangover.

Implementation of the ethics program can be seen as the dawn of the *company's ethical culture*, which we discussed in the "Unethical Behavior within Organizations" section of Chapter 4. And the *ethics training and communication* component of the program should be evolved and repeated over and over again to reinforce the ethical business practices in everyone's mind. Unfortunately, only 29% of respondents to the Nordic Business Ethics Survey reported that this happens in their organization.

TABLE 10.2
The Seven Group Essentials of the Volkswagen Group (This Information Belongs to Volkswagen and Is Being Used to Demonstrate How Reputable Companies Are Guiding Their Employees)

Group Essential
I. We take on responsibility for the environment and society
II. We are honest and speak up when something is wrong
III. We break new ground
IV. We live diversity
V. We are proud of the work we do
VI. We not me
VII. We keep our word

Source: [3].

In addition, it is important to measure the state of workplace ethics regularly in order to monitor changes in the company's ethical culture. But the measurement interval should be long enough—for example, a year—because the time constants associated with changes in employees' ethical behavior are quite long; too short a measurement interval cannot show a possible improvement or deterioration. A questionnaire similar to the one used in the Nordic Business Ethics Survey could serve as an effective instrument even within a single (fairly large) organization. And a mobile user interface combined with an AI-based analysis, reporting, and action-proposal tool would make the thoughtful questionnaire convenient and affordable to use. In this way, companies could monitor the state of workplace ethics regularly and swiftly without the help of (costly) business consultants.

Toward Certification?

Business Ethics Standard

In Chapter 2, we recognized an analogy between "quality management" and "business ethics" within companies. Since the quality sector has its well-proven ISO 9000 Quality Management certificate [4], could business ethics also have a similar universally recognized certification? By the end of 2022, approximately a million companies worldwide had applied for and received ISO 9000 registration—mostly ISO 9001, which belongs to the same *family* of standards—in an effort to show that their procedures and processes are in line with global quality requirements (*Source*: www.statista.com). A similar business ethics certificate would increase the confidence of investors and customers in a company and make it a more dependable employer to join as a new employee; simply, the company's moral credibility could therefore be improved. To conclude, the purpose of business ethics certification would be to provide a globally recognized standard, which could be used for the external assurance of the state of business ethics on an individual company basis.

On the other hand, the ISO 9000 certification process is considered to be laborious and costly [4]. And it is obvious that if a uniform standard of business ethics existed, its corresponding auditing process with necessary preparations would have similar undesirable characteristics. But a more serious hurdle is the considerable variation between international work cultures. Generally speaking, the global work cultures of the West, East, and South tend to be distinct. This could potentially lead to a *weak* standard, as it has to adapt to a wide range of regional and organizational traditions and work cultures. In this regard, the ISO 9000 family of quality standards has a more unified context; quality management does not go into the ethical values of employees. Anyway, such a business ethics certification is certainly worth pursuing, but it will take time before one becomes available.

Conformity to Requirements

In numerous companies, quality is considered pragmatically as *conformity to requirements*, not as an absolute quantity to aim for. So, their cost-effective products and services must be "good enough"—nothing more, nothing less. Consequently, business ethics should also conform to specific requirements. "Absolute ethics," like Immanuel Kant's Deontology (Chapter 3), is not a realistic objective in business and engineering at large. And concrete requirements, whose compliance can be explicitly monitored, could be placed on such items as:

- Humane working conditions and practices
- Transparency of decision-making in relation to employees
- Corruption-free organization
- No misconducts that lead to settlement costs or fines
- Organization without workplace harassment
- Continuously developing business ethics practices

These requirements aim to support the values of the organization. For example, in the Nordic Business Ethics Survey, 34% of the respondents had observed *discrimination or bullying* in their company during the past year; an acceptable requirement for this unethical item should be an "asymptotic zero." And not surprisingly, such business ethics requirements constitute a kind of "detailed code of ethics." In addition to the requirements, companies should have well-defined procedures for recording possible *deviations* as cases; as well as for case pre-processing, corrective actions, monitoring of handled cases and learning from them; and also for the continuous development of this vital process.

Recognizing deviations, such as workplace harassment, should not be left only to potential whistleblowers, but managers ought to address such significant ethical issues in departmental meetings and annual supervisor–subordinate discussions. And it would be recommended to introduce an Anti-Harassment Pledge (see model below) in workplaces, which everyone should sign during a collective session.

> **Anti-Harassment Pledge:** *We are committed to never engaging in any form of workplace harassment. If we become aware of such behavior in our workplace, we will not hesitate to report it.*

In order to ensure the success of the Anti-Harassment Pledge, it is very important to get employees involved in its implementation.

WHISTLEBLOWERS AND ETHICS AMBASSADORS

Whistleblowing Dilemma

In the style of Shakespeare's Hamlet: "To be *fair*, or to be *loyal*, that is the question." This ethical dilemma is often pondered when an employee has observed unethical conduct at his/her workplace. And we know that "to be loyal" is a more prevalent norm in collectivist cultures (e.g., Chinese and Japanese) than in individualistic cultures (e.g., American and German). On the other hand, "to be fair" is pivotal in countries like Australia and Norway. Dungan et al. state expressively in their scholarly article: "Whistleblowing, reporting another person's unethical behavior to a third party, represents an ethicist's version of optical illusion" [5]. For a moment, this thought-provoking statement was enigmatic to the authors of this book. How would you interpret it?

Overall, 64% of respondents to the Nordic Business Ethics Survey confirmed that they did *not* intervene when they witnessed unethical or illegal conduct in their workplace. For different reasons, they did not blow the whistle. And Ovaska does not recall a single case of whistleblowing during his 13-year industrial career in Finland and Kentucky. It seems that the virtue of fairness often loses competition with the virtue of loyalty. In fact, an identified whistleblower can face considerable backlash and even retaliation from colleagues or management.

Who Blows the Whistle?

What motivates some employees to blow the whistle in the first place, while others might hesitate in a similar situation? That is another Shakespearean question. Dungan et al. suggested three factors influencing the decision of whether or not to blow the whistle [5]:

1. *Personal* factors; proactive versus reserved
2. *Cultural* factors; inter- versus in-dependent
3. *Situational* factors; protected versus threatened

They also pointed out that "the few employee demographic factors that correlate with higher rates of whistleblowing include increased tenure of employment at the company, increased pay, increased education, and being male." And among personality traits, people high in extroversion are more likely to blow the whistle. Whistleblowers are typically individuals with personality traits that support nonconformity, and they are often seen as independent persons who do not care what others think of their doings. Moreover, organizational encouragement for whistleblowing, dissemination of knowledge about the options for reporting unethical conduct, and trusted safety measures to protect whistleblowers from retaliation all increase the likelihood of blowing the whistle when needed. But variation in cultural norms that emphasize or de-emphasize loyalty affects the likelihood of whistleblowing. Sometimes, employees who would, in principle, be ready to blow the whistle do not act because they believe the particular unethical conduct is none of their business or that whistleblowing would not change anything or, even worse, that such conduct is common practice.

These research-based findings parted the impenetrable curtain of whistleblowing for practitioners. Regrettably, only 17% of respondents to the Nordic Business Ethics Survey confirmed the availability of a web-based whistleblowing tool that enables *anonymous* (opposite to acknowledged) reporting, which would be desired in most cases. Perhaps there should exist a whistleblowing app today. Finally, Dungan et al. concluded: "To motivate a broader swath of individuals toward whistleblowing, organizations might focus on building the kind of community that values constructive dissent while maintaining group loyalty" [5]. In general, whistleblowing can be seen as both a right and an obligation; and it should definitely be used when the unethical incident is "big enough." But it is up to the individual to decide what is big enough for him/her; that is a Shakespearean decision relying on one's free will. On the other hand, whistleblowing is not needed in organizations with a high ethical culture.

I'm No Informant!

In Chapter 9, in the section "New Avenue for Experienced Engineers," we had an opportunity to ask the anonymous electrical engineer why he had not adopted value-based ethics as a guideline in practicing his profession. Later, we also asked his opinion about whistleblowing. Right after hearing the word "whistleblowing," he became upset. He compared whistleblowers in the corporate environment to civilian informants in the former East Germany (officially the German Democratic Republic), an authoritarian society in the middle of the Cold War. The feared Ministry for State Security, commonly known as the *Stasi* ("Staatssicherheit"), had a dense civilian informant network throughout society. People were distrustful and constantly on their toes because the secret Stasi informant could be, for instance, a colleague, neighbor, friend, or even a relative. The overall atmosphere was obviously paranoid. And our anonymous engineer saw the ad-hoc whistleblowers as the *equals* of East German civilian informants. "They destroy the working atmosphere and ruin collaboration, and thus degrade the efficiency of the organization as well as the wellbeing of its members. No! Instead, the problems of unethical conduct within organizations must be addressed *preventively* through company-wide ethics programs and moral coaching." Well, this was just his personal, strong opinion.

The issue of whistleblowing is, indeed, multifaceted. From a pragmatic point of view, whistleblowing in some cases appears *heroic*, while in other cases it appears *reprehensible* [5]. But does it make sense to compare corporate whistleblowers to those civilian informants in the past East Germany? There is a fundamental difference between them. What is that difference? We leave these questions for the readers to ponder.

Promoting Ethical Culture on the Frontline

Network of Ethics Ambassadors

Ethics ambassadors form a peer network of volunteer employees who dedicate their developing proficiency in business ethics (in a secondary role) to the benefit of their organization and its members. They act as an interface between employees and (senior) management and promote an ethical culture based on common core values in the organization [6]. Nevertheless, they are not part of an ethics office but individuals who are considered as "one of us." Ethics ambassadors are typically located in different business units, geographical locations, and at different levels of the corporate hierarchy. And they should preferably represent diverse demographic groups. Furthermore, they are sometimes called Ethics Advisors or Ethics Representatives. Such roles are common, especially in *large* multinational companies with thousands of employees. Ethics ambassadors have up-to-date information about their company's ethics program, and they can help evolve the program, reach employees from within the organization, and promote the company's ethical culture at the local level—especially among colleagues working nearby. And it is important that the members of the ethics ambassador network meet from time to time—either virtually or physically—to share their experiences and learn from each other. A study conducted by the Institute of Business Ethics stated, "ethics ambassadors fulfil an important role in raising awareness among employees, in their division or stage of each of the main building blocks of an ethics programme" [6].

We can also see the ethics ambassadors as a potential quasi-anonymous channel for whistleblowing. This thinking is based on the ethics ambassadors' *duty of confidentiality*. The ethics ambassador can record the confidential case and give it an anonymous identifier, forward it for appropriate processing, and finally inform the whistleblower of the outcome. In this way, the potential whistleblower has a chance to discuss his/her concerns in peace with the ethics ambassador before the whistle is actually blown. In several cases, such a pragmatic discussion may resolve the unethical concern, and no whistleblowing is needed. In this way, every quasi-anonymous whistleblowing case is surely handled in the organization, none is ignored without a justified explanation.

Furthermore, in recognition of exemplary volunteer contributions in promoting an ethical culture, corporations could establish an *Outstanding Ethics Ambassador Award*, which includes a plaque and a monetary reward of maybe $1,000.

The Everyday Life of an Ethics Ambassador

But what about the reality in the "trenches"? To better understand the role and tasks of ethics ambassadors, we got an opportunity to interview Sami-Seppo Ovaska, a member of the Stora Enso Ethics Ambassador network. That illuminating interview is transcribed in the vignette below.

INTERVIEW WITH AN ETHICS AMBASSADOR

Hi Sami, good to meet you!

Q1: Where do you work?

A1: I currently work at Stora Enso as a specialist in the Innovation, Research and Development organization of the Packaging Materials Division, in Finland. Stora Enso is a multinational company that provides renewable products in packaging, biomaterials, and wooden constructions. The division I represent focuses on fiber-based premium packaging materials.

Q2: What is your education?

A2: I hold a D.Sc. degree in chemical engineering from Lappeenranta University of Technology. My major was paper technology, and my doctoral research focused on investigating the potential of pigment-filled dispersion coatings to act as grease barriers in paperboard-based food packaging materials. Other topics I worked with were inkjet printing and the converting of packaging materials. During my doctoral education, I acted as a project manager in several industrial joint research projects and supervised multiple M.Sc. theses, and I also gave lectures related to paper-making and biomaterials in undergraduate courses.

Q3: What is your primary job function?

A3: I am a responsible researcher in the Packaging Laboratory, and my day-to-day work includes plenty of collaboration and communication with our internal customers all over the world. I am responsible for planning analytical testing, reporting findings, and following the development of packaging and analytical sectors. My role involves providing expert support for our technical customer service, sales organization, and multiple internal projects in the fields of packaging materials, barriers, package converting, and end-use performance of packages.

Q4: How did you become an ethics ambassador?

A4: I simply noticed an announcement on my employer's intranet about ethics ambassador training. After discussing this opportunity with my manager, I enrolled in the training program. The training consisted of distance lectures and some homework, and after four months, I received my diploma.

Q5: What have you done so far in this role?

A5: Ethics ambassadors are volunteers who promote ethics and shared values in their organization. The ethics ambassadors have short quarterly meetings and sometimes attend training sessions about specific topics, such as the ethics of AI. I have also participated in many regular meetings of different teams in our Research Center to share information about the ethics ambassador network, maintain ethical awareness, and introduce special ethical themes. However, I would say that my most important task has been, and will continue to be, acting as a low-threshold interlocutor in situations where a colleague has noticed something suspicious or suspects that they have been mistreated. In such situations, I may act as an intermediary between the colleague and the Ethics and Compliance team so that the colleague can anonymously seek professional advice for his/her problem.

Q6: How much of your working time is spent acting as an ethics ambassador?

A6: There is no specific time allocated for acting as an ethics ambassador, and the time spent varies a lot throughout the year. My best estimate is that I spend a total of 15–20 hours per year on these activities. In terms of time, it is not a major effort to act as an ethics ambassador.

Q7: Why do you want to be in that particular role?

A7: I think the driving force in my case has been my personal values, which motivated me to enroll in the ethics ambassador training. After a long engineering career in academia and industry, I realized that I was not that familiar with the social sciences [note: behavioral ethics is a subfield of the social sciences], and I wanted to develop professionally in that area as well. My experiences so far have been very positive, and I believe that as an organization we can achieve better outcomes if we do keep ethics in mind. It is also worth mentioning that there are no significant contradictions between my personal values and those of my employer.

Q8: How does your employer benefit from the ethics ambassador program?

A8: The ethics ambassador program can benefit employers in many ways. Such programs support building internal company culture, for example, by increasing inclusion. Improved compliance is definitely one aspect—it is essential that employees are aware of their ethical expectations and obligations. A better understanding of shared values may increase employee commitment and a feeling of togetherness, which in turn may promote ethical leadership. Ideally, ethics ambassador programs also support diversity and can help to reduce the occurrence of ethical misconduct. It must be kept in mind that the legal and financial consequences of misconduct can be considerable, and whistleblowers may prevent unethical actions from going too far.

Q9: Would you recommend this kind of role to other engineers, as well?

A9: It is highly recommended, indeed. Ethics is an essential but often forgotten component of engineering. It is not uncommon for engineers to be faced with ethical dilemmas. These dilemmas can be related to product safety, environmental impact, social responsibility, or basic interpersonal relationships, for instance. The training itself provides better skills to understand human behavior and also some tools to deal with sensitive situations. However, as an ethics ambassador, one may end up in a situation where they are a person of trust for their colleague. Therefore, the ambassador must be capable of being discreet, fully credible, and appreciative. It is also of great importance that one is genuinely ready to speak up if they notice any kind of violation of the company's rules or values. It is strongly about helping create a culture of ethical behavior and promoting trust within the work community. However, an ethics ambassador with an engineering background is not a legal expert, and thus it is essential to stay as unbiased as possible and to refrain from giving any legal advice even in the most blatant or demanding situations.

Thanks a lot, Sami, and have a good one!

The cost-effective network of ethics ambassadors can have a significant value for companies; but it may be hard to find motivated *engineers* with the right personality traits for this demanding secondary role. A potential candidate should ideally be friendly, a patient listener, and a trusted and empathetic person who copes with difficult situations calmly—and, of course, has a genuine interest in business ethics. Could you perhaps be interested?

MEASURABLE BENEFITS: UNETHICAL VERSUS ETHICAL PRACTICES

The pursuit of various benefits is a notable driver behind human behavior, and gaining benefits also represents a key motivator behind business actions in companies. Unfortunately, the means by which these benefits are obtained can sometimes be unethical or even illegal. In the following, we first discuss the strive for short-term gains through deception or other forms of unethical conduct in the company world. Finally, the rational way of adhering to ethical business practices is suggested as a viable choice that pays off in the long run.

The Temptation of Short-Term Gains

Why do organizations end up using unethical practices? Ultimately, because there are people who make decisions and carry out the corresponding actions; and people may be weak to certain temptations, both in private and professional life. As highlighted in Chapter 1, there are always *individuals* behind unethical actions—and organizations, as structures, are made up of individuals. Some of those can be engineers or engineering managers. Unethical business practices are essentially a sign of poor organizational self-discipline, rooted in the company's unhealthy values. Therefore, the company's values must be reformed before it is possible to "do what's right." And for that, we need a CEO who is committed to the implementation of value-based ethical business practices.

Unethical Business Conduct

Bribery has been used, among other things, to secure large infrastructure contracts across different sectors, such as mobile communications networks, power plants, and large security systems. In such cases, bribes can amount to millions of dollars and may be paid to politicians or government officials in the target country. Although these bribes are substantial, the obtained gains are usually significantly higher: more contracts in the future, increased revenue, higher operating income, generous bonuses for the management team, and so forth. But the benefits may be short-lived because someone—such as a watchful competitor or an investigative journalist—might suddenly blow the whistle.

Cartels between a few companies in the same industry can be used to regulate the pricing of specific products or services. Together, these competing companies may secretly agree on a relatively high price level for a certain type of product or agree on certain market areas or specific segments where they do not compete with each other. This is naturally illegal. And the gains may include increased revenue, higher operating income, and notable bonuses for the management team. But also, these benefits may be short-lived because someone—such as an envious competitor or competition authorities—can discover the wicked cartel.

Unethical employment practices, incredibly low wages, and poor working conditions, especially in developing countries, are a brutal way to minimize the manufacturing costs of various products. Compared to manufacturing in countries that regulate fair employment and working conditions (e.g., Myanmar versus Denmark), the gains can include significantly lower manufacturing costs, higher overall sales, increased operating income, and substantial bonuses for the management team. Nevertheless, news media or human rights organizations may sooner or later discover these types of unethical business practices.

Sometimes companies include *fraudulent functionality* in their products, such as the infamous "defeat device" we discussed in Chapter 8. There is an illusion that since the detailed operations of embedded software cannot be observed externally, it is possible to build "software-cheating" functions, for instance, on communications, control, and automation systems, with minor risk of being caught. While the risk may be relatively low, it is never zero. Because of this low risk, the obtained benefits can easily last for a medium term, say a few years. And they can include competitive advantage, increased sales, prestigious quality and other awards, higher operating income, substantial bonuses for the management team, and so on. But it is always possible that either some research laboratory or a vengeful ex-employee will make the software-cheating feature public.

The last example we present here is not straightforward at all and is related to *environmental ethics*. Leading companies and even societies (or politicians) sometimes keep the *total* environmental impact of certain equipment or constructs under the table. For example, "clean energy production" might be promoted without openly acknowledging the existence of unsustainable stages between manufacturing (incl. mining and extraction) and decommissioning (incl. waste management) of equipment and associated constructs. In these cases, the versatile chain of benefits can include overly optimistic CO_2 and other pollutant figures, relaxed construction licensing procedures, increasing sales, high operating incomes, and bonuses for management teams—even recognition for politicians. Such cases are hard to tackle because they can involve multiple companies, government officials, and legislators. Whistleblowers could, in principle, be intrepid scientists, investigative journalists, or environmental groups, but their voices would be relatively low in this kind of case. Even if whistleblowing were acknowledged and taken seriously, it would take a long time to run down such a large-scale activity if there exists no realistic alternative. So, it seems to make sense to join the "august club" of these stakeholders, right? ("It was the famous Don Quixote who fought against *windmills*.")

The Iceberg Hypothesis

An unethical way of doing business can be financially beneficial in the short-term, maybe even longer. But companies that engage in such practices are wasting their resources to cover up these secrets. And if their scandalous unethical practices are disclosed, a spiral of declining profits may begin. The disclosed cases of bribery, cartels, unethical employment practices, and product fraud can be so big and impudent that they develop into scandals that we read about in the news media. However, the majority of scandalous cases likely remain undisclosed and we never hear about them. This train of thought led us to create the following hypothesis.

Hypothesis: The disclosed unethical business scandals are only the tip of the iceberg compared to the number of undisclosed, scandalous business ethics actions. (Typically, about 10% of an iceberg is visible above sea level.)

Even if this hypothesis appears to be true, it cannot be proved true or false—because the undisclosed cases are, indeed, unknown to us. Thus, we can only speculate. And we can only hope that companies do not use the de facto message of this hypothesis for *calculated risk-taking* when contemplating an unethical alternative to conducting their business; this is clearly a value issue.

Adherence to Ethical Business Practices Pays Off

After reflecting on the rather disturbing message of the previous section, it is good to note that 48% of the respondents to the Nordic Business Ethics survey (Table 10.1) reported that doing ethical business is *already* financially rewarding—and that was in 2020. And they do consider it financially viable even though the implementation of an ethics program, a stimulating platform for a company's ethical culture, is a considerable expense. On the other hand, ethics programs should be seen as *investments* rather than expenses because they lead toward more ethical business practices. Furthermore, the possible benefits of ethical business practices and company cultures include the following:

- No deterioration of the corporate image or reputation, which would affect sales, recruitment, and investor confidence
- No court costs, settlement expenses, or fines related to unethical conduct
- No bribes or penalties related to corruption
- No theft or embezzlement in the workplace
- Lower staff turnover (less recruitment and orientation expenses)
- No harassment in the workplace (fewer days of absence)
- Better wellbeing at work (improved work efficiency)
- No work-related ethical hangovers (more content and loyal employees)

Depending on the degree of the company's unethical past, the realizable benefits can range from moderate to huge. But the bottom line is that an ethics program pays off *only* when it is seriously implemented with genuine commitment from senior management. In fact, the single most important factor in achieving ethical behavior in an organization is the commitment of senior management to this noble goal [7]. And in the long run, it provides benefits to shareholders ($++), customers (product credibility), the organization itself (improved work efficiency), and employees (moral wellbeing). So, everyone is going to be a winner!

CRISIS AND ETHICS

In general, when people are faced with a crisis, they enter an alert mode. This holistic mode can significantly alter their usual thinking and behavior patterns. However, the outwardly visible behavior can vary greatly from person to person; some remain calm, while others can become anxious—it depends on the individual's personality traits and his/her life situation.

In this section, we discuss organizational crises that eventually turn into personal crises for individual employees. Such personal crises, triggered by stressful uncertainties within organizations, can push some employees to a sort of "fight-or-flight" state, where "fight" could mean unethical behavior toward colleagues and

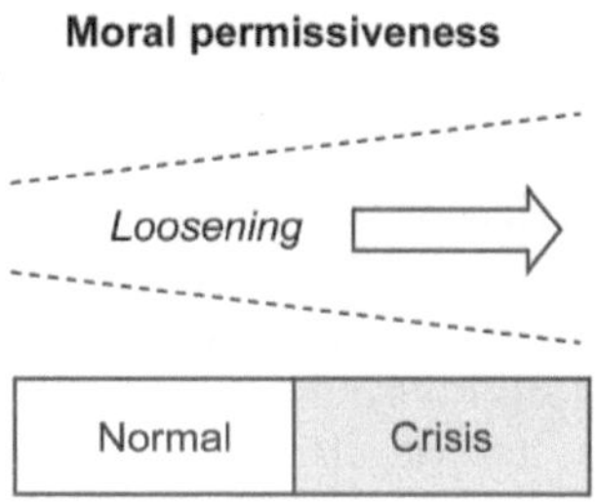

FIGURE 10.1 Employees' moral permissiveness is gradually loosened when organizational conditions change from normal to crisis. Note that the linguistic variables "Normal" and "Crisis" represent varying degrees of organizational conditions; "Normal" has a decreasing degree, while "Crisis" has an increasing degree, from left to right.

management and "flight" would mean leaving the company as soon as a new job is found. This dual-fork behavior focuses on self-preservation, and these employees may respond with idiosyncratic patterns meant to address only their personal priorities and concerns.

Jurkiewicz writes in her encyclopedia article: "Specifically, crisis situations present unique and prime opportunities for unethical behavior to occur, both opportunistic unethicality or that perceived as justifiable given the situation" [8]. And later she continues: "Ethics in practice appears to be a superfluity to which people attend after their basic needs are satisfied." These empirical observations set the stage for increasingly unethical conduct. Figure 10.1 illustrates the related *moral permissiveness* of employees that loosens as organizational conditions shift from normal toward crisis.

Organization in Crisis

Organizations can face a variety of crises, but here, we only consider those economic crises that lead to operational downsizing and layoffs of many employees. In such a challenging situation, employees can be classified into two categories: *survivors* and *victims*. To successfully recover from the ongoing economic crisis and sustain long-term viability for their company, managers have to re-engage surviving employees and minimize the potential for retaliatory behavior from victims of layoffs [9]. A few of these victims may feel morally justified in retaliating against management (and thereby grant themselves an associated moral self-license), believing it was unfair to receive a "layoff ticket." Thus, they may think that responsible managers should pay for their wrongdoings—leaning on misunderstood fairness [10] (pp. 81–83).

The commendable study by McDewitt et al. [9] identified perceptual differences between the survivors and victims concerning the following factors:

- Management's responsiveness to employee needs during downsizing
- Fairness of layoff decisions
- Possible favoritism in layoff decisions

- Optimism of employees regarding the future of the organization
- Employees' nervousness about job security

In general, survivors experienced these factors in a more positive way than victims, which is intuitively understandable. These are, indeed, sensitive factors that deserve careful consideration. When it comes to layoff decisions, observed favoritism can have long-lasting effects even among survivors, who may adopt questionable means to stay on the boss's "favorite list." For instance, to position oneself as the boss's close golf partner might provide job security in certain contexts—it can even be seen as a form of "employment insurance."

At this point, a tempting question could be: is it possible to decrease one's risk of being laid off by unethical means—such as smearing colleagues—during the weeks or months when downsizing and layoffs are being considered in employee consultation discussions? Or is the increased moral permissiveness (Figure 10.1) merely a harmful artifact under these conditions?

Ethics as a Resilient Backbone

Smearing of colleagues is of no positive use—period. If one's behavior becomes unethical or retaliatory, the only outcome is that both colleagues and management have the opportunity to observe that the individual possesses a somewhat difficult personality. This observation becomes a negative merit. Difficult employees are more likely to be laid off during downsizing. So, try to stay calm and avoid goofing around on the unethical trail. Ethical behavior forms a resilient backbone for employees who await layoff decisions in fear, as well as for the victims after the layoff tickets are handed out.

When organizations are downsizing and laying off engineers, it is not only your knowledge capital that maximizes the chances to retain your job. In addition, you are expected to be a team player, at least a moderate communicator, and it surely helps if you are a committed and hard-working employee [11]. But none of these qualities can be notably improved during the short period of employee consultation discussions. On the other hand, all these qualities can be cultivated over time through continuous self-improvement and professional development; *you are continuously responsible for your career.*

Furthermore, be prepared for organizational crises, as there is no such thing as a 100% secure job. Even major segments of the high-tech industry may disappear unexpectedly. For example, the technological evolution had dramatic consequences for Nokia Corporation, the leading manufacturer of cell phones, prior to the touchscreen smartphone era. Nokia's (later Microsoft Mobile's) mass layoffs left thousands of engineers unemployed [11]. Their expertise included mobile radio, communications, microelectronics, electronics manufacturing, and real-time software engineering. Therefore, it is necessary to brainstorm back-up career paths every now and then—an organizational crisis may be lurking around the corner. But the good news is that engineers are educated and experienced *problem solvers*, and thus they can aim to apply the well-proven "engineering process" (it provides a framework) to solving their own employment problems, too [12].

Moreover, during an organizational crisis, the company may be close to a marginally stable state, and the possible retaliatory actions of victim employees (transient-like excitations to the overall system) could even undermine the stability of this *complex system*, called an organization. Of course, there are also other factors that affect the stability of an organization in crisis. We can confidently conclude that the increased moral permissiveness of employees in an organizational crisis must be seen as a harmful artifact.

Finally, it is important to highlight that organizational crises can go beyond personal crises and develop into regional crises. This situation may arise when large factories or major business units shut down, leading to sharply rising unemployment rates and significant drops in tax revenues. These regional crises often bring with them numerous ethical considerations, as well.

SUMMARY

Business ethics programs establish an indispensable basis for the flourishing of an ethical culture in companies. But they are only effective when top management is genuinely behind them and leading by sincere example. In this chapter, we presented a general-purpose ethics program and its multi-phase implementation. This could serve as an initial template for companies since they usually tailor their ethics programs to their special needs and constraints.

Whistleblowing keeps on having a contradictory tone in organizations. In any case, it is sometimes needed to stop unethical behavior before that reaches scandalous proportions or leads to serious consequences for a harassed employee, for instance. It is, indeed, an important "air bag" for maintaining ethical standards in organizations.

In addition, ethics ambassadors form a cost-effective network of volunteers to advance business ethics in their companies. Because they are considered "one of us," it should be pretty easy for fellow employees to turn to them with sensitive ethical questions and concerns. They form a bridge connecting employees, ethical guidelines, and management, which promotes an organization's positive ethical atmosphere.

Bribery, cartels, unethical employment practices, and fraudulent functionality in products may bring substantial monetary benefits to companies engaging in such unethical practices. However, these "dirty" benefits often have a short lifespan—or do they? Once discovered, they lead to settling fees, penalties, and fines, erasing the unethically obtained gains. Fortunately, nearly half of the respondents to the Nordic Business Ethics Survey (Table 10.1) stated that ethical business practices are already financially rewarding. Adherence to ethical business practices not only pays off but also ensures long-term sustainability.

Companies face economic crises from time to time, which appears to be a kind of business rule. When a crisis strikes, leaders have to make agonizing choices under pressure. Simultaneously, employees become alert and distressed over the uncertainty of their job security—are they going to be fired? This stressful situation triggers crisis-management mechanisms in individuals, often leading to a loosening of moral permissiveness. However, such a loosened moral permissiveness should not be used toward colleagues or management because it can exacerbate the situation, leading *only* to negative outcomes. Especially in a crisis situation, it is good if employees have firm ethical norms that they can rely on. Moreover, ethics should be seen as a resilient backbone that supports the suffering organization.

Now, after ten chapters exploring various facets of Ethics for Engineers with a unique and pragmatic emphasis, we are getting ready to present the forward-looking epilogue. It highlights the humane perspective on ethics and morality, while also delving into the exciting opportunities that future AI chatbots will provide for the engineering community and their academic educators in the context of professional ethics.

CLASS EXERCISES

1. Create a mind map of this chapter.
2. For a moment, the authors of this book considered the following sentence of Dungan et al. enigmatic: "Whistleblowing, reporting another person's unethical behavior to a third party, represents an ethicist's version of optical illusion" [5]. How would you interpret the meaning of that metaphorical sentence? You can use one of the AI chatbots as your assistant.
3. Our fellow electrical engineer compared *whistleblowers* in the corporate environment to *civilian informants* in the former East Germany—as discussed above. Is it fair to make such a juxtaposition? Justify your answer.
4. Could you be interested in volunteering as an *ethics ambassador* in your (future) engineering organization? Why or why not?

   Class discussion: Compare the answers of different students and try to explain the dominance of "yes" or "no" answers.
5. There is an iceberg hypothesis in the section of "The Temptation of Short-Term Gains," which states: *The disclosed unethical business scandals are only the tip of the iceberg compared to the number of undisclosed, scandalous business ethics actions.* Unfortunately, this particular hypothesis cannot be proved to be true or false.

   Class debate: The instructor moderates a prepared debate between two volunteer teams. First, the class is divided into three teams; one of the teams is favoring the positive option ("the hypothesis is true"), the other is against it ("it is false"), and the third team evaluates the arguments presented. What are the objective conclusions of the evaluation team?
6. Based on your own experiences and observations, do you think that people's moral permissiveness (Figure 10.1) really loosens when they face crises in their personal or professional lives? If your answer is "yes," please give an example of such. And if it is "no," try to justify your opinion. Furthermore, how is this seen from the *Deontology* perspective?
7. Consider the VW diesel emissions scandal discussed in Chapters 1 and 8. Assume that one of the software developers—who programmed the defeat device code—disclosed that unethical conduct to the public a couple of years after its implementation. What consequences could he/she face after bravely blowing the whistle?

Class discussion: Compare the answers of different students; and answer the question, did this whistleblowing make sense from the *Utilitarian* point of view?

8. In the section on "Crisis and Ethics," we speculated that to position oneself as the boss's close golf partner might provide job security in certain contexts—it can even be seen as a form of "employment insurance."

   Class discussion: Does this kind of favoritism really exist in practice? Is it good or bad from the *Virtue Theoretical* viewpoint? Could you consider such an attractive "employment insurance" yourself?

REFERENCES

1. "Nordic Business Ethics Survey 2020: A study of Nordic employees' perception of ethics at work." NBE. Accessed: Jul. 22, 2024. [Online]. Available: https://www.nordicbusinessethics.com/wp-content/uploads/2020/10/Nordic-Business-Ethics-Survey-2020.pdf
2. M. Kaptein, "The effectiveness of ethics programs: The role of scope, composition, and sequence," *Journal of Business Ethics*, vol. 132, pp. 415–431, Dec. 2015, doi: 10.1007/s10551-014-2296-3
3. "Our Code." Volkswagen. Accessed: Jul. 22, 2024. [Online]. Available: https://www.volkswagen-group.com/en/integrity-and-compliance-15705
4. T. H. Stevenson and F. C. Barnes, "Fourteen years of ISO 9000: Impact, criticisms, costs, and benefits," *Business Horizons*, vol. 44, no. 3, pp. 45–51, 2001, doi: 10.1016/S0007-6813(01)80034-3
5. J. Dungan, A. Waytz, and L. Young, "The psychology of whistleblowing," *Current Opinion in Psychology*, vol. 6, pp. 129–133, Dec. 2015, doi: 10.1016/j.copsyc.2015.07.005
6. "Ethics Ambassadors: Promoting ethics on the front line." IBE. Accessed: Jul. 22, 2024. [Online]. Available: https://www.ibe.org.uk/resource/ethics-ambassadors-promoting-ethics-on-the-front-line.html
7. R. W. Clement, "Just how unethical is American business?" *Business Horizons*, vol. 49, no. 4, pp. 313–327, 2006, doi: 10.1016/j.bushor.2005.11.003
8. C. L. Jurkiewicz, "Ethics and crisis management," in *Global Encyclopedia of Public Administration, Public Policy, and Governance*, A. Farazmand, Ed., Cham, Switzerland: Springer, 2017, doi: 10.1007/978-3-319-31816-5_2749-1
9. R. McDevitt, C. Giapponi, and D. M. Houston, "Organizational downsizing during an economic crisis: Survivors' and victims' perspectives," *Organization Management Journal*, vol. 10, no. 4, pp. 227–239, 2013, doi: 10.1080/15416518.2013.859057
10. S. Bok, *Lying: Moral Choice in Public and Private Life*. New York, NY: Pantheon Books, 1978.
11. S. J. Ovaska, "Managing your career in a dynamic environment," *IEEE Potentials*, vol. 37, no. 3, pp. 24–26, 2018, doi: 10.1109/MPOT.2017.2764512
12. J. Treichler, "A team sport," *IEEE Potentials*, vol. 37, no. 3, pp. 43–45, 2018, doi: 10.1109/MPOT.2018.2794660

Epilogue

Authoring a textbook of this breadth and depth is actually like docking a ship; both require knowledge, meticulousness, and dedication. When we began our exciting book project, we had four initial objectives:

1. The text should cover both moral theoretical and behavioral ethics perspectives.
2. It should be based on day-to-day micro-level cases from the lives of practicing engineers, supplemented with macro-level cases.
3. It should provide pragmatic guidance for individual engineers and their organizations to move toward value-based ethics.
4. The language should be colloquial to make the book an enjoyable and accessible read for engineering students.

Consequently, the completed text, which includes 29 demonstrative vignettes, 87 class exercises, and an insightful interview with an ethics ambassador, will serve as a pedagogically sound learning companion for courses in engineering ethics and related topics. This unique text strikes a balance between research-based findings (with over 40 scholarly references) and real-world experiences (featuring an Appendix by an industry executive), making it a valuable resource for both students and educators alike. It is literally something *What Every Engineer Should Know* (the title of our Book Series).

Before we continue with this Epilogue, it is still useful to place engineering ethics in context. Engineering ethics is a *subcategory* of both professional and business ethics, and it involves using moral reasoning to identify the responsibilities and standards that ought to guide engineers in their professional practice. For example, the Volkswagen "Dieselgate" (Chapter 8) clearly demonstrates that engineering ethics is, indeed, a subcategory within the broader domain of business ethics.

ETHICS IS HUMANE

A Central Ethical Question

Ethical people behave ethically, which is natural. But people who behave ethically are not necessarily ethical. All one needs to *behave* ethically is basically agency and self-discipline. But *being* ethical is deeper; one should be empathetic and possess the two core virtues of compassion and respect that make the person humane. Outwardly, the end result in both cases may appear the same, but internally they are far from each other. Thus, while ethics is humane, ethical behavior may not be; instead, in some extreme cases, it may even be a matter of cold calculation.

Here, we are facing a central ethical question: Is ethics primarily about *actions* or about *character*? In other words, which one of these questions is more fundamental: What should I do? or Who should I be? They are related, of course. People who perform right actions tend to have good characters, and people with good characters tend to perform right actions.

This absorbing and important question has received renewed attention in recent years and has roots reaching back to the Ancient Greek philosophers. Aristotle, for instance, raises the issue in his Nicomachean Ethics (the term "Nicomachean" is believed to refer to Aristotle's son, Nicomachus), where he states [1]:

> Actions, then, are called just and temperate when they are such as the just or the temperate man would do; but it is not the man who does these that is just and temperate, but the man who also does them as just and temperate men do them. It is well said, then, that it is by doing just acts that the just man is produced, and by doing temperate acts the temperate man; without doing these no one would have even a prospect of becoming good.

And the following is a philosopher's interpretation of Aristotle's words for practitioners: What Aristotle means is that a person could perform virtuous *acts* without possessing a virtuous *character*. Of course, if she performs them long enough, she may become good. But we should not mistake a virtuous action for a virtuous person.

As you may recall, the above text of Aristotle was already presented in Chapter 3. We wanted to repeat it here to emphasize the extensive timeline of ethical considerations—spanning over two millennia.

Why Does It Matter?

In business ethics, we are primarily interested in ethical business practices and "doing what's right," rather than the underlying character of the employees within organizations. In general, behaving ethically in accordance with the company's code of conduct would seem to be sufficient.

But is there any such role in engineering organizations where it would be beneficial for a person to *be* ethical—not only to behave ethically? Sure, the role of an ethics ambassador (Chapter 10) is a good example—and it is certainly not the only one. A competent ethics ambassador embodies humane qualities and ethical character, making him/her more approachable. This, in turn, enhances the ambassador's credibility during consultations with employees and management.

Furthermore, the "being–behaving" distinction applies also to organizations. A company might establish an ethics office as a facade to appear concerned about moral responsibility; such practice falls under the category of "ethics-washing." In that case, we would not say that the company is an ethical one. But if the company is genuinely committed to ethical business practices, we would consider it ethical. And this really makes a difference in the long run.

Interestingly, a somewhat similar difference can be identified between human intelligence and artificial intelligence. While humans have the powerful traits of agency and emotional depth, *all* AI systems do lack these. This fundamental distinction was highlighted nearly two decades ago in the memorable panel address, "Computer-Based Intelligence: Where Is It Going?" by Tony Martinez of Brigham Young University [2]. It was presented at an IEEE Systems, Man, and Cybernetics workshop on adaptive and learning systems.

GENERATIVE ARTIFICIAL INTELLIGENCE IN PROFESSIONAL ETHICS

EDUCATIONAL CHALLENGES

Generative AI chatbots, like OpenAI's *ChatGPT*, Microsoft's *Copilot*, and Google's *Gemini*, are a timely discussion topic also in engineering education, and this discussion is likely to continue for years [3]. Currently, some teaching faculty may still be skeptical and frustrated, worrying about how to keep their students away from these "problematic" tools. However, AI chatbots are continuously developing and penetrating all segments of society; that is an inevitable fact. Therefore, instead of wasting time and resources on trying to ban their use or detecting inappropriate usage, we could adopt an open-minded point of view. So, how about focusing on developing *novel* teaching, learning, tutoring, and evaluation methods for the next generation of courses? This will require revolutionary (opposite to evolutionary) solutions; but isn't it the only rational alternative? Fortunately, the whole academic world is struggling with the same challenges. Hence, there is a large pool of international contributors around this substantial transformation.

In fact, the ongoing generative AI boom has similarities to the microprocessor boom that began in the early 1970s [4]. As we know, microprocessors are now ubiquitous. But their early years were also full of skepticism and frustration. Of course, the world is much more technologically advanced now than it was in those historical days.

WHAT COULD GENERATIVE AI BE USED FOR?

Next, we explore two thought-provoking examples of the potential use of AI chatbots in professional ethics applications. They are intended to inspire readers to innovate and use generative AI applications that would advance the ultimate quest to "do what's right."

My Casual Ethics Advisor

When evaluating the current capabilities of AI chatbots in creating ethical analyses of practical scenarios, we took one of the vignettes from Chapter 7 (Box 7.5) and asked *Copilot* to analyze Walter's behavior in that story; the outcome is below. We suggest readers to compare it to the corresponding Analysis and Discussion sections in Chapter 7.

COPILOT'S ETHICAL ASSESSMENT OF THE CASE "THIS IS EXTORTION!"

Our Question to Copilot: *Was Walter's behavior in the following story ethical* [see Box 7.5]?

Copilot's Reply: The ethical evaluation of Walter's behavior in this story involves considering several factors. Let's break it down:

1. Erich's Transparency:
   - Erich openly expressed his desire for a multi-year assignment to a foreign subsidiary. He was forthright about considering leaving the company if such an opportunity wasn't available. While this honesty is commendable, it also set the stage for the conflict.
2. Walter's Reaction:
   - Walter's response was strong and emotional. He accused Erich of extorting him, which escalated the situation. His anger and labeling of Erich's behavior as "revealing a great deal about his personality" were unprofessional.
   - Walter's reaction might be seen as an overreaction, especially considering their good relationship and shared experiences.
3. Erich's Revelation:
   - Erich revealed that Walter had used a similar tactic in the past to secure a temporary assignment in Japan. This revelation put Walter on the defensive.
   - While Erich's intention may have been to level the playing field, it also damaged trust between them.
4. Walter's Denial and Lying:
   - Walter vehemently denied Erich's claim, calling it absurd and asserting that he would never engage in such behavior.
   - However, when confronted with Egon's account, Walter's credibility suffered. His lie eroded trust further.
5. Impact on Their Relationship:
   - The conflict left both Walter and Erich upset and strained their previously strong relationship.
   - Trust was weakened, and their interactions changed significantly.

Ethical Assessment:

- Walter's initial accusation of extortion was inappropriate and unprofessional.
- Erich's revelation about Walter's past behavior added complexity to the situation.
- Walter's denial and subsequent lie damaged trust and integrity.
- Both individuals contributed to the breakdown in their relationship.

In summary, Walter's behavior was ethically questionable due to his emotional outburst, denial, and dishonesty. Erich's transparency, while commendable, also led to unintended consequences. Ultimately, both should have handled the situation more professionally and transparently.

"Free tier Copilot with GPT-4 model" (23 February 2024)

Well, a moral philosopher would comment that the above assessment is superficial and vague—these are certainly valid points. However, the assessment is also pragmatic. And it may encourage practitioners to utilize AI chatbots as "savvy advisors" who can provide counseling in ethically challenging issues. Until now, no casual ethics advisor has been available for everyday use. Although such savvy advisors could be appreciated by practicing engineers, they are not in competition with moral philosophers and their profound theoretical perspective. On the other hand, in the long run, AI chatbots might represent some kind of sparring with behavioral ethicists. Still, their roles remain complementary rather than competitive.

Maybe you could also try such a savvy advisor. But beware: the more we rely on AI to point out our failings, the less capable we become of recognizing those failings ourselves—and *we* are responsible for ourselves.

Ethical Conduct Tracker

The previous example presented a low-threshold application of generative AI to an individual-level ethics issue, while the following blue-sky application looks at an intriguing application idea at the organizational level. In Chapter 10, we introduced the hypothesis: *The disclosed unethical business scandals are only the tip of the iceberg, compared to the number of undisclosed, scandalous business ethics actions.* As we noted, this cannot be proved to be true or false, because we cannot know the undisclosed actions.

But instead of aiming to prove it one way or the other, we could develop an "ethical conduct tracker," which would assess the *likelihood* of undisclosed, scandalous business actions within the company of interest. And that would be based on large-scale data analytics [5] combined with generative artificial intelligence [6].

For this, we would need time-series of various business and financial data from the company *and* its closest competitors (for benchmarking) over a lengthy period of time. And this data is supplemented by related market data, news headlines, social media posts, and so forth. The data matrices could include, for example, the following business metrics:

- Revenue
- Operating income
- Share price
- Paid dividend per share
- Bonuses for senior management
- Secured major contracts

- Number of employees
- Downsizing and layoffs
- Substantial recruitment of new workforce
- Significant awards and prizes

It is intuitive that scandalous, unethical business actions leave traces in business metrics and public data, because such actions are typically aimed at obtaining financial benefits for the company or its management team, in one way or another. Therefore, we believe this type of *big data* must contain valuable information tidbits that, when put together, can provide insightful knowledge about the state of ethical conduct. The discovered knowledge can be fed into an AI chatbot, which could eventually produce a rationalized likelihood—much like weather forecasting—regarding undisclosed, scandalous business actions within the company. Such an end result is sketched in the following example.

> **Example:** The AI chatbot could report that there are undisclosed, scandalous business actions within the company with 70% confidence. And this conclusion would be justified by the particular patterns and anomalies detected in the analyzed data. In addition, a temporal trend of the evolving confidence could be provided.

Our early idea has the potential to enhance corporate governance and applied ethical standards. It could, for instance, end up being a valuable tool for large corporations to proactively monitor and improve the ethical conduct of their global subsidiaries. We believe that such an intelligent tracker would make a significant impact on corporate ethics and transparency. But it should definitely serve as an *assistant*, not as the master.

The first ethical conduct trackers might become available within the next 3–5 years. For this extensive research and development effort, expertise is needed from diverse fields, such as business ethics, business law, artificial intelligence, data analytics, systems engineering, organizational psychology, as well as corporate finance and management. And the research and development could be carried out as an international collaboration effort between several universities and piloting corporations.

To conclude, large-scale data analytics, combined with generative artificial intelligence, offers innovative potential even in the rather traditional field of business ethics. Hopefully, in the intermediate-term future, the tracking results of possible scandalous, unethical conduct could gradually be made available for various stakeholders to follow online. This voluntary transparency would be beneficial in maintaining a high level of *trust* in corporations, especially if the ethical conduct tracker were a service run by a non-profit organization. Moreover, the ethical conduct tracker could be a potential component of the *business ethics certification* envisioned in Chapter 10.

Nevertheless, more important than these emerging tools is the evolution of the behavior and character of individual employees and managers—as well as their companies—toward higher ethical standards. Thus, the current trend of professionals and corporations adopting higher ethical norms is gratifying to observe.

FIGURE E.1 So near and yet so far: A segment of the Mödlareuth "Little Berlin" Wall that once divided East and West Germany. (Photo (21 May 2023) courtesy of Seppo Ovaska.)

THE FINAL LINE

Remember, *every ethical hangover is a chance*. How does Figure E.1 relate to this?

REFERENCES

1. Aristotle, *Nicomachean Ethics*. Book II, Section 4, translated by W. D. Ross, originally published 350 BCE. Cambridge, MA: MIT. Accessed: Aug. 16, 2024. [Online]. Available: https://classics.mit.edu/Aristotle/nicomachaen.2.ii.html
2. T. Martinez, "Computer-based intelligence: Where is it going?" Opening Address in Panel Discussion, *IEEE Mountain Workshop on Adaptive and Learning Systems*, Logan, UT, July 26, 2006.
3. S. S. Gill et al., "Transformative effects of ChatGPT on modern education: Emerging era of AI chatbots," *Internet of Things and Cyber-Physical Systems*, vol. 4, pp. 19–23, 2024, doi: 10.1016/j.iotcps.2023.06.002. [Online]. Available: [Open Access]
4. S. J. Ovaska, "If I were a student again: My next choice," *IEEE Potentials*, vol. 37, no. 3, pp. 41–42, 2018, doi: 10.1109/MPOT.2017.2734941
5. S. M. Srinivasan and P. A. Laplante, *What Every Engineer Should Know About Data-Driven Analytics*. Boca Raton, FL: CRC Press, 2023.
6. G. Yenduri et al., "GPT (Generative Pre-trained Transformer)—A comprehensive review on enabling technologies, potential applications, emerging challenges, and future directions," *IEEE Access*, vol. 12, pp. 54608–54649, Apr. 2024, doi: 10.1109/ACCESS.2024.3389497. [Online]. Available: [Open Access].

Appendix A: Ethics and Compliance Program in a Major International Company

BACKGROUND

Stora Enso's operations started some 1,000 years ago, and the company is thereby one of the oldest in the world. During the centuries, the operations and business models have changed several times, and the ability to change, depending on trends and markets conditions, is a major strength in the company's essence.

Today, Stora Enso is a major global player in the circular bioeconomy. A leading provider of renewable products in *packaging*, *biomaterials*, and *wooden construction*, and one of the largest private forest owners in the world. Stora Enso creates value with low-carbon and recyclable fiber-based products, through which the company supports its customers in meeting the demand for renewable sustainable products. Stora Enso has approximately 20,000 employees, and the sales were €9.4 billion in 2023.

Operating in the circular bioeconomy with production of renewable and sustainable products, and managing huge forest assets, makes it fundamentally important to run the operations, including the whole value chain, in a sustainable and ethically responsible way. Acting without integrity and in a non-responsible way would totally undermine the trust from investors, customers, and other stakeholders. A sustainable business model is hence an important part of the company's strategy.

The company is exposed to several risks, both in the areas of sustainability and business ethics. The running of forest operations, in the Nordic countries, Latin America, and Asia involves human rights and environmental challenges. Moreover, the whole supply chain can entail exposure to similar risks. Stora Enso also operates in locations including high-risk emerging markets, which offer good business opportunities but may also cause serious risks relating to topics, such as corruption and fraud. The structure of the market where Stora Enso operates, with rather few big players and high entry barriers, also exposes the company to risks related to competition laws.

CHANGE OF THE LANDSCAPE

Compliance with laws and other requirements has always been important for companies in the heavy industry sector. In the old times, it was mostly to avoid major operational disruptions, fines, or administrative penalties.

During the last decades, the regulatory regime has changed dramatically. The UK Bribery Act and the U.S. Foreign Corrupt Practices Act are examples of legislation that governments and authorities have shaped to combat corruption. Similar examples can be given in the area of competition law where national and intergovernmental authorities, such as the European Commission, have far reaching executive power to enforce the legislation. These laws and practices place high demands on the controlling mechanisms of companies, but they also help to build accountability and trust among stakeholders.

Furthermore, we can today see a totally different and increased focus and interest on environmental, social, and governance-related topics from customers, investors, employees, and other stakeholders. In today's media landscape, different kinds of unacceptable behavior can be shared with millions of people on social media with just a few taps on a smartphone, resulting in severe adverse media exposure.

These factors have made compliance one of the major business challenges for any multinational company. What used to be a nice-to-have "hygiene factor" is today a license to operate and thereby an important part of company sustainability.

WHY DO GOOD PEOPLE DO BAD THINGS?

Almost every day, when we open the newspaper, we can read about another company that has been caught red-handed in bribing foreign public officials or engaging in a cartel with its competitors. The illegal behavior might have been going on for years, and many employees might have known about it without interfering. Eventually, a whistleblower speaks up or the authorities knock on the door in a dawn raid.

Surprisingly often, these companies have rolled out excellent steering documents with impressive statements of the importance of strict compliance with laws and regulations, employed hundreds of compliance officers and implemented systems to control the employees. Still, company representatives, even top management, make intentional or careless decisions to cut corners, to achieve short-term gains in an even more competitive environment.

The reasons why this happens, repeatedly, are naturally different for different companies and individuals. But the most common explanation is what behavioral scientists have referred to as "ethical blindness." Even though the individuals are well educated and know what is right and wrong, the root cause for these scandals is not "bad apples" in the organization but rather the context in which the bad behavior has emerged. A toxic environment is often composed by different kinds of *pressure*. For example, peer pressure, time pressure, or pressure from the management. Very often, it is also combined with *fear*—fear to raise an objection, fear not to be part of the group, or fear to be a disloyal and troublesome colleague.

Hence, as the foundation of any compliance program, only focusing its efforts on strict compliance with statements in steering documents, controlling employees, and having the view that the compliance program exists only to avoid risks and bad behavior—even though important issues—will not fundamentally change the behavior of employees and lead to a healthy company culture. Rather, the company should aim to establish a *value-driven* culture where people are guided by a common moral compass when faced with difficult decisions, act with integrity, and speak up against misconduct or unethical behavior.

In addition, it is important to underline that the ethics and compliance program is not only there to protect against risks. And compliance is not only about following the laws and regulations; it is also about acting in an ethical and responsible way, even when there is no legal obligation to do so. In today's complex business world, the importance of ethical conduct cannot be overstated. While profit and success are certainly important, they must be achieved in a manner that is consistent with ethical principles and values. To believe otherwise is to fundamentally misunderstand what is important in life.

Many companies today are taking proactive steps to align their business practices with their values and principles and to respond to the expectations of their stakeholders. For instance, hundreds of companies, Stora Enso included, decided to exit or suspend their operations in Russia after its invasion of Ukraine, despite the significant financial losses they incurred. They did so because they wanted to show their support for human rights and international law, and to avoid being associated with a regime that violated these.

THE STORA ENSO ETHICS AND COMPLIANCE STRATEGY

The way Stora Enso has aimed to enforce its policies and guidelines in the compliance area, as well as change to a more value-driven company, is formulated in the *Ethics and Compliance Strategy*. The current strategy, along with its annually updated action plan, explains how the Ethics and Compliance function supports the realization of and safeguards Stora Enso's overall *Purpose*—"Do good for people and the planet; replace fossil-based materials with renewable solutions"; and *Values*—"Lead" and "Do What's Right," as well as to contribute to the fulfillment of the company's business strategy and objectives.

To realize this holistic objective, ethics and compliance has adopted the following purpose: "Building Trust through Integrity." As said earlier, sound and ethical business behavior is not only a risk mitigator. In today's business context, it is also one of the most important enablers and value creator for business. Stora Enso's long-term success depends on the trust and approval of all our stakeholders, including investors, business partners, customers, and our current and future employees.

Trust results from doing the right thing continuously and systematically. Building a *Culture of Integrity* will make sure all of us want to do the right thing, not because it is prescribed in a steering document or because we are monitored, but because we believe in and do share our purpose and values. In this respect, ethics and compliance can play a very important role, contributing to Stora Enso's business strategy and becoming a forerunner also in terms of its ethical business practices.

Fostering a culture of integrity requires constant activities throughout the company, activities that need to be repeated year after year, because the company memory is rather short and new employees are recruited every year. The four key areas below are the building blocks of our culture of integrity. For each of these areas, action plans with key performance indicators (KPIs) are developed, executed, and monitored on a yearly basis.

Key I: ***Risk Assessment and Compliance.*** We must identify potential risks in advance, analyze them, and take precautionary steps to mitigate risks. Without such risk management, a cost-efficient compliance program cannot be realized. The need for effective risk management is further underlined by increased business complexity as well as new and strengthened legislation in several business areas.

To understand and mitigate our risks, we should further analyze the extensive data that can be extracted from different processes within the company. We need to draw conclusions from compliance investigations and further liaise with other expert functions in the company to better understand the business and monitor selected risk groups. Also, we need to build a strong ethics and compliance organization with the ability to act swiftly in a fast-paced environment, as well as take necessary precautions in challenging business environments.

We should also liaise and cooperate with business operations and other service functions as their trusted strategic partner, including to actively collaborate with business owners in assessing and understanding the ever-evolving regulatory landscape, including new legislation, reputational risks, and geo-political challenges. Also, by supporting the business in customer or supplier audits, and by showing the company's ethical stand in relation to competitors.

Even though Stora Enso has the ambition to be a value-driven organization, where the focus is on ethics rather than compliance, policies, and guidelines, their effective processes are instrumental. Policies and guidelines serve as important handrails for how to navigate based on our values in difficult business circumstances.

The Stora Enso Code

Our code of conduct, the Stora Enso Code, is a single set of values for all our employees, a guideline that explains our approach to *ethical business practices*, *human and labor rights*, as well as *environmental values*. It guides our work and is applied wherever we operate.

The Business Practice Policy is designed to provide employees with more detailed guidelines on how to comply with the high-level principles set out in the Stora Enso Code and to provide a framework for what we consider responsible conduct in our daily business activities.

Effective processes increase efficiency and make it easier to "Do What's Right." At the same time, solid processes for investigations and a centralized approach with fairness in consequences support compliance.

Digitalization and innovation are important enablers to deliver our compliance program and associated systems in a smart way. We aim to utilize existing digital infrastructures and adopt new solutions that improve the quality and efficiency of our processes. We collaborate with IT, Finance, and other relevant functions to achieve synergies when possible. Examples of our digitalization and innovation accomplishments include robotic process automation for trainings, an AI-based chatbot, fraud detection algorithms, and so on.

Key II: ***Value-Based Leadership.*** Organizational cultures are built over time as beliefs and values are established and reinforced. A clear and unfailing tone from the top is the fundamental prerequisite. Such ambitions need to be supported with consistent communication, leadership on all levels and allies for change throughout the organization. Commitment requires openly discussing beliefs and values to ensure all employees are attuned to the ethical dimensions and challenges of their work.

To build an organization with a solid ethical foundation, top management must actively promote a culture of compliance and *zero tolerance* for unethical behavior. Effective tone and sound leadership will build trust in the organization and exemplify by behavior the type of conduct expected. At the same time, it must be clear what is considered unacceptable and what are the consequences of compliance breaches. Leadership needs to be strengthened—not only on the very top but on all levels of management.

Key III: ***Ethical Dialogue.*** Clear communication and frequent training are fundamental for increased awareness and value setting among employees and business partners. Ethics and compliance is realizing this objective by regularly training *all* Stora Enso employees in different topics, via e-learning or face-to-face, by communicating important messages and topics, and by further developing Stora Enso's network of Ethics Ambassadors.

Inclusion and diversity help us in creating a successful business environment but can also help us in dealing with ethical challenges. Diverse perspectives help teams avoid the problem of "tunnel vision," that is, where decisions are based on a too narrow outlook or problems addressed with obsolete solutions. Diverse teams are more likely to avoid ethical blindness, unconscious biases, and preconceptions. In this area, a close co-operation with the HR organization is vital.

Key IV: ***Speak up—Listen up.*** Creating a culture of openness and inclusion, where people are encouraged to give their viewpoints and voice their concerns, is fundamental for a healthy and competitive work environment. The main reason behind many corporate scandals is that group pressure among the employees and pressure from the management have created a culture of cover-up and silencing, rather than promoting a healthy culture of speaking up.

Creating this culture of openness is a dual responsibility. All managers in Stora Enso should promote diversity of thought and it is essential that a speak-up culture is matched by a willingness to also listen up. Employees should be encouraged to be active team members and support their managers in creating team cultures based on open feedback.

Furthermore, we must all learn to speak up for one another and avoid being bystanders. We need to further encourage Stora Enso's employees and business partners to get involved and "Do What's Right" by reporting suspected misconduct, illegal behavior, or breaches of the Stora Enso Code.

Stora Enso recognizes that a critical aspect of its compliance program is the establishment of a culture that promotes *prevention*, *detection*, and *resolution* of unethical behavior and conduct that does not comply to law or policies and procedures. Hence, we enforce the Stora Enso Code, as well as underlying policies and guidelines, by investigating any reports of misconduct in an unbiased manner. Where infringements are identified, actions will be taken to prevent similar incidents in the future.

THE ETHICS AND COMPLIANCE PROGRAM GOING FORWARD

This book Appendix started by describing the fundamental change Stora Enso has undergone during the centuries. The same thing will in the future be said about the Ethics and Compliance program. The program needs to evolve, in pace with changed regulatory requirements and changed business environment, in order to stay relevant and provide good business support. During the last years, the area of responsibility for the ethics and compliance function in most companies, also in Stora Enso, has expanded year by year.

One example is the new tasks and responsibilities in the area of sustainability, where we can see an avalanche of new initiatives from both national legislators and the European Union (EU), such as the Corporate Sustainability Reporting Directive (CSRD), the Corporate Sustainability Due Diligence Directive (CSDDD), and the EU Deforestation Directive (EUDR).

These are examples of "soft law" becoming "hard law" in such a way that previously non-binding instruments are transformed into legally binding obligations, thereby also moving parts of the responsibility to Ethics and Compliance. As a result of this legislation, companies will have a legal obligation to address issues, such as forced labor, child labor, and environmental damage in their value chains. Companies are expected to have policies and practices in place to identify and mitigate these risks, such as conducting regular audits, engaging with suppliers and customers on sustainability issues, and adopting responsible value chain practices.

These changes go hand in hand with the fact that investors and other stakeholders are increasingly looking at Environmental, Social, and Governance (ESG) factors when making business decisions. Companies are expected to have strong ESG policies and practices in place, especially in areas such as climate change, human rights, and corporate governance. This includes reporting on ESG metrics, setting ambitious sustainability goals, and disclosing the risks and opportunities associated with ESG factors.

Another area that will fundamentally change the working practices of any ethics and compliance functions is artificial intelligence (AI). AI can help ethics and compliance teams to identify and assess potential risks that may arise from business operations. AI-powered tools can analyze vast amounts of data from multiple sources, including news articles, social media, and financial reports, to identify potential risks and provide insights into emerging trends and issues.

Generative AI can also be used to monitor and detect potential violations of laws, regulations, or company policies. This includes monitoring of employee communications, financial transactions, and other business operations. AI-powered tools can identify patterns of behavior that may indicate wrongdoing, such as unusual transaction patterns or suspicious employee behavior.

SUMMARY

When talking about ethics and compliance and motivating why companies at all should spend time and money on extensive compliance programs, we tend to focus only on avoiding risks. Risks of losing customers. Risks of having to pay fines. Risks of getting bad publicity. And this is quite natural, because nearly every day we read about companies running into problems, and where the consequences are so severe that they might threaten the existence of the whole company. However, what if we instead viewed ethics and compliance as an *opportunity*? How about instead focusing on the "compliance business case?"

All in all, acting according to the Stora Enso Code and the company values is key to achieving Stora Enso's strategic objectives and to maintain the trust of our stakeholders. This is truly an opportunity for us, but also an obligation, because as a big and important player in the circular bioeconomy, Stora Enso has not only a possibility, but even an obligation, to "Do Good for the People and the Planet."

Pontus Selderman
Senior Vice President, Ethics and Compliance
Stora Enso, Sweden

Appendix B: IEEE Code of Ethics

We, the members of the IEEE, in recognition of the importance of our technologies in affecting the quality of life throughout the world, and in accepting a personal obligation to our profession, its members and the communities we serve, do hereby commit ourselves to the highest ethical and professional conduct and agree:

I. **To uphold the highest standards of integrity, responsible behavior, and ethical conduct in professional activities.**

1. to hold paramount the safety, health, and welfare of the public, to strive to comply with ethical design and sustainable development practices, to protect the privacy of others, and to disclose promptly factors that might endanger the public or the environment;
2. to improve the understanding by individuals and society of the capabilities and societal implications of conventional and emerging technologies, including intelligent systems;
3. to avoid real or perceived conflicts of interest whenever possible, and to disclose them to affected parties when they do exist;
4. to avoid unlawful conduct in professional activities, and to reject bribery in all its forms;
5. to seek, accept, and offer honest criticism of technical work, to acknowledge and correct errors, to be honest and realistic in stating claims or estimates based on available data, and to credit properly the contributions of others;
6. to maintain and improve our technical competence and to undertake technological tasks for others only if qualified by training or experience, or after full disclosure of pertinent limitations;

II. **To treat all persons fairly and with respect, to not engage in harassment or discrimination, and to avoid injuring others.**

7. to treat all persons fairly and with respect, and to not engage in discrimination based on characteristics such as race, religion, gender, disability, age, national origin, sexual orientation, gender identity, or gender expression;
8. to not engage in harassment of any kind, including sexual harassment or bullying behavior;
9. to avoid injuring others, their property, reputation, or employment by false or malicious actions, rumors or any other verbal or physical abuses;

III. **To strive to ensure this code is upheld by colleagues and co-workers.**

10. to support colleagues and co-workers in following this code of ethics, to strive to ensure the code is upheld, and to not retaliate against individuals reporting a violation.

Adopted by the IEEE Board of Directors and incorporating revisions through June 2020.

Index

Note: **Bold** and *italic* page numbers refer to **tables** and *figures*, respectively.

F

G

H

I

J

K

L

M

N

O

P